Wavelet Analysis in Civil Engineering

Pranesh Chatterjee
Tata Steel, Netherlands

CRC Press
Taylor & Francis Group
Boca Raton London New York

CRC Press is an imprint of the
Taylor & Francis Group, an **informa** business

CRC Press
Taylor & Francis Group
6000 Broken Sound Parkway NW, Suite 300
Boca Raton, FL 33487-2742

First issued in paperback 2017

ISBN-13: 978-1-4822-1055-2 (hbk)
ISBN-13: 978-1-138-89395-5 (pbk)

Visit the Taylor & Francis Web site at
http://www.taylorandfrancis.com

and the CRC Press Web site at
http://www.crcpress.com

Contents

Preface

When I started my doctoral research, my supervisor introduced me to the concept of wavelets. Initially, I was quite suspicious about the term *wavelets*. In time I came to learn that a wavelet is a powerful signal processing tool and can do some amazing things – extract the features hidden in a signal, for example. I became fond of wavelets and used the wavelet analytic technique for most of my research work. However, I realized that this particular topic was not easy to digest. More specifically, it was difficult to find suitable books outlining the application of wavelets in engineering. Since then I had in mind a latent desire to write such a book from my experience in this area that would be beneficial to interested postgraduate students and researchers. The opportunity came when I visited the stall of CRC Press at the 15th World Conference on Earthquake Engineering in 2012 in Lisbon, Portugal. I expressed my wish to write a book on applications of wavelets in civil engineering, and was approached soon thereafter when we all decided to go ahead.

I am greatly indebted to Professor Biswajit Basu and Professor Mira Mitra, who shared their valuable experiences with me and never hesitated to carry out in-depth discussions at times on the topic. I am grateful to my wife, Tanima, who has been a constant source of inspiration in my work, as well as to my two children, Sraman and Soham, who spent quality time with me to break up the monotony of writing. Last but not the least, sincere thanks are due to my friend Dr Debashish Bhattacharjee, who always encouraged me to remain strong in the face of daunting tasks.

Pranesh Chatterjee
Delft, the Netherlands

MATLAB® is a registered trademark of The MathWorks, Inc. For product information, please contact:

The MathWorks, Inc.

3 Apple Hill Drive

Natick, MA 01760-2098 USA

Tel: 508-647-7000

Fax: 508-647-7001

E-mail: info@mathworks.com
Web: www.mathworks.com

Acknowledgments

The author gratefully acknowledges the support of the Slovenian National Building and Civil Engineering Institute, ZAG, for provision of data from the SiWIM Bridge Weigh-in-Motion system (www.siwim.si) used in this book.

About the Author

Pranesh Chatterjee, PhD, earned undergraduate and postgraduate degrees in civil engineering and a doctorate in engineering from Jadavpur University, India. His main research focus during his doctoral study was in the field of structural dynamics. During this research work he extensively used the wavelet-based analytical technique to formulate various problems in soil–structure–fluid interaction analyses. In later stages, Dr Chatterjee took up a postdoctoral fellowship in the Structural Mechanics section at Katholieke Universiteit te Leuven in Belgium and then was selected as a prestigious Pierse Newman Scholar at the University College Dublin in Ireland. After spending a considerable amount of time in academics and participating in a number of interesting research works that resulted in several journal and conference publications, he decided to move to industry. Since then he has worked in different fields and currently works as manager of the plasticity and tribology group of Tata Steel Europe in the Netherlands. Dr Chatterjee has always been active in research and its publication.

Introduction

The main objective of Chapter 1 is to introduce to readers the concept and utility of wavelet transform. It begins with a brief history of wavelets referring to earlier works completed by renowned researchers, followed by an explanation of the Fourier transform. The chapter also shows the advantages of the wavelet transform over the Fourier transform through simple examples, and establishes the efficiency of the wavelet transform in signal processing and related areas. Chapter 2 first describes the discretization of ground motions using wavelet coefficients. Later, it explains the formulation of equations of motion for a single-degree-of-freedom system in the wavelet domain, and subsequently the same is used to build the formulation for multi-degree-of-freedom systems. The systems are assumed to behave in a linear fashion in this chapter. The wavelet domain formulation of equilibrium conditions of the systems and their solutions in terms of the expected largest peak responses form the basis of the technique of wavelet-based formulation for later chapters. Chapter 3 focuses on two distinct problems. The first is to explain how to characterize nonstationary ground motion using statistical functionals of wavelet coefficients of seismic accelerations. The second is to develop the formulation of a linear single-degree-of-freedom system based on the technique as described in Chapter 2 to obtain the pseudospectral acceleration response of the system. The relevant results are also presented at the end. Chapter 4 shows stepwise development of the formulation of a structure idealized as a linear multi-degree-of-freedom system in terms of wavelet coefficients. The formulation considers dynamic soil–structure interaction effects and also dynamic soil–fluid–structure interaction effects for specific cases. A number of interesting results are also presented at the end of the chapter, including a comparison between wavelet-based analysis and time history simulation. Chapter 5 describes the wavelet domain formulation of a nonlinear single-degree-of-freedom system. In this case, the nonlinearity is introduced into the system using a Duffing oscillator, and the solution is obtained through the perturbation method. Chapter 6 introduces the concept of probability in the wavelet-based theoretical formulation of a nonlinear two-degree-of-freedom system. The nonlinearity is considered through a bilinear hysteretic spring, and the probability conditions are introduced depending on the position of the spring with respect to its yield displacement condition. The analysis is supplemented with some numerical results. In the last chapter (Chapter 7), focus is on diverse applications to make readers aware of the use of wavelets in these areas. For this purpose, three different cases are discussed. The first one is related to the analysis of signals from bridge vibrations to identify axles of vehicles passing over the bridge. The second example explains the basic concept and formulation of stiffness degradation using a physical model. Thereafter, the chapter focuses on using a numerical technique to obtain the results of a degraded model (stiffness degradation through formation of cracks) and then compares the wavelet-based analysis of the results obtained from linear and nonlinear models. The third example is related to soil–structure–soil interaction. In this example, the wavelet analytic technique is used to obtain the results at the base of a structure considering dynamic soil–structure interaction. Subsequently, the forces, shears and moments thus obtained at the base of the model are applied at the supporting soil surface and a three-dimensional numerical model of this structure–soil interaction problem is used to obtain a nonstationary response within the soil domain.

Chapter 1

Introduction to wavelets

1.1 HISTORY OF WAVELETS

In 1807, Joseph Fourier developed a method that could represent a signal with a series of coefficients based on an analysis function. The mathematical basis of Fourier transform led to the development of wavelet transform in later stages. Alfred Haar, in his PhD thesis in 1910 [1], was the first person to mention wavelets. The superiority of Haar basis function (varying on scale/frequency) to Fourier basis functions was found by Paul Levy in 1930. The area of wavelets has been extensively studied and developed from the 1970s. Jean Morlet, who was working as a geophysical engineer in an oil company, wanted to analyse a signal that had a lot of information in time as well as frequency. With the intention of having a good frequency resolution at low-frequency components, he could have used narrow-band short-time Fourier transform. On the other hand, in order to obtain good time resolution corresponding to high-frequency components, he could also have opted for broad-band short-time Fourier transform. However, aiming for one meant losing the other, and Morlet did not want to lose any of this information. Morlet used a smooth Gaussian window (representing a cosine waveform) and chose to compress this window in time to get a higher-frequency component or spread it to capture a lower-frequency component. In fact, he shifted these functions in time to cover the whole time range of interest. Thus, his analysis consisted of two most important criteria – dilation (in frequency) and translation (in time) – which form the basis of wavelet transform. Morlet called his wavelets 'wavelets of constant shape', which later was changed by other researchers only to 'wavelets'. J.O. Stromberg [2] and later Yves Meyer [3] constructed orthonormal wavelet basis functions. Alex Grossmann and Jean Morlet in 1981 [4] derived the transformation method to decompose a signal into wavelet coefficients and reconstruct the original signal again. In 1986, Stephen Mallat and Yves Meyer developed multiresolution analysis using wavelets [3, 5, 6], which later in 1998 was used by Daubechies to construct her own family of wavelets. In

1996, Daubechies [7] gave a nice, concise description of the development of wavelets starting from Morlet through Grossmann, Mallat, Meyer, Battle and Lemarié to Coifman, from the 1970s through mid-1990s. A pool of academicians, including pure mathematicians, engineers, theoretical and applied physicists, geophysical specialists and many others, have developed various kinds of wavelets to serve specific or general purposes as and when needed. Thus, though initiated mainly by the mathematicians, wavelets have gained immense popularity in all fields of applied sciences and engineering due to their unique time–frequency localization feature. It is due to this unique property that the wavelet transform has proved its ability (and reliability) in analysing nonstationary processes to reveal apparently hidden information that no other tool could provide. The application areas are wide, e.g. geophysics, astrophysics, image analysis, signal processing, telecommunication systems, speech processing, denoising, image compression and so forth. The wavelets have been applied analysing vibration signals. Some special techniques like discrete and fast wavelet transforms have been developed for this purpose. Before going into the discussion on wavelet analytic technique any further, it would be wise to review the basic theory on Fourier transform at this point.

1.2 FOURIER TRANSFORM

Most of the single-valued functions may be written as the summation of a series of harmonic functions within a desired range. This series is termed Fourier series. The concept of such a series has already been used by Daniel Bernoulli in connection to solving problems of string vibrations. However, it was Joseph Fourier, the French mathematician, who did a systematic study on Fourier series for the first time. Fourier series has found many applications in the fields of heat conduction, acoustics, vibration analysis, etc. The Fourier series for a function $f(x)$ in the interval $\alpha < x < \alpha + 2\pi$ is written as follows:

$$f(x) = \frac{a_0}{2} + \sum_{n=1}^{\infty} a_n \cos(nx) + \sum_{n=1}^{\infty} b_n \sin(nx) \tag{1.1}$$

where

$$a_0 = \frac{1}{\pi} \int_{\alpha}^{\alpha+2\pi} f(x)dx \tag{1.2}$$

$$a_n = \frac{1}{\pi} \int_{\alpha}^{\alpha+2\pi} f(x)\cos(nx)dx \tag{1.3}$$

$$b_n = \frac{1}{\pi} \int_{\alpha}^{\alpha+2\pi} f(x)\sin(nx)dx \tag{1.4}$$

The readers should note that once the value of α is chosen as zero, the interval becomes $0 < x < 2\pi$, and on choosing $\alpha = -\pi$, the range becomes $-\pi < x < \pi$. Thus, Fourier series can actually represent any periodic (and also nonperiodic) function as the sum of simple sine and cosine waves. This idea forms the basis of Fourier transform (FT), which is an extension of the Fourier series. In Fourier transform, the period of the function may extend to infinity. The Fourier transform retrieves the frequency content of a signal. It decomposes a signal into orthogonal trigonometric basis functions. The Fourier transform $\hat{X}(\omega)$ of a continuous function $x(t)$ is defined in the following equation:

$$\hat{X}(\omega) = \frac{1}{\sqrt{2\pi}} \int_{-\infty}^{\infty} x(t)e^{-i\omega t}dt \tag{1.5}$$

In the above equation, the term $\hat{X}(\omega)$ gives the global frequency distribution of the time-dependent original signal $x(t)$. The original signal $x(t)$ can be further reconstructed using the inverse Fourier transform as defined below:

$$x(t) = \frac{1}{\sqrt{2\pi}} \int_{-\infty}^{\infty} \hat{X}(\omega)e^{i\omega t}d\omega \tag{1.6}$$

The following Dirichlet conditions must be satisfied for Fourier transform and its reconstruction:

1. The time function $x(t)$ and its Fourier transform $\hat{X}(\omega)$ must be single-valued and piece-wise continuous.
2. The integral $\int_{-\infty}^{\infty} |x(t)|dt$ must exist that insists that if $|\omega| \to \infty$, $\hat{X}(\omega) \to \infty$.
3. The functions $x(t)$ and $\hat{X}(\omega)$ have upper and lower bounds (however, this is not a necessary condition).

In case of signals obtained from experiments, they are discrete in nature as they are sampled at N discrete time points with a sampling time of, say, Δt. These signals are analysed in the frequency domain using the concept of discrete Fourier transform (DFT), $\hat{X}_{DFT}(\omega)$, as defined below:

$$\hat{X}_{DFT}(f_n) = \frac{1}{N}\sum_{n=0}^{N-1} x(n)e^{-i2\pi n\Delta t} \tag{1.7}$$

It may be seen from Equation (1.7) that the DFT may be evaluated at discrete frequencies $f_n = \frac{n}{N\Delta t}$, where $n = 0, 1, 2, \ldots, N - 1$. The inverse DFT, as shown below, may be used to get back the original discrete time signal.

$$x(n) = \frac{1}{\Delta t}\sum_{f_n=0}^{\frac{N-1}{N\Delta t}} \hat{X}_{DFT}(f_n)e^{i2\pi f_n n\Delta t} \tag{1.8}$$

It may be noted here that the $N\Delta t$ in the equation above denotes the time length of the signal. The discrete Fourier transform computation requires evaluation of real and imaginary parts separately; thus, $2N^2$ numbers of operations would be required. So, DFT works quite well when the signal length is short. If the signal becomes large with numerous discrete time points, DFT could become very tedious. The idea of fast Fourier transform (FFT) is developed, which is computationally more efficient in such cases because the FFT algorithm works on signals that must have as many samples as the power of 2 (i.e. 2^m samples). The FFT is much faster because it uses the results from previous computations and thereby reduces the number of operations required. It utilizes the periodicity and symmetry of trigonometric functions to compute the transform with approximately $N\log N$ numbers of operations.

If the time-dependent function $x(t)$ in Equation (1.6) has only one frequency, the corresponding frequency spectrum, $\hat{X}(\omega)$, is a Dirac delta function. So, if the frequency spectrum has only one frequency, say, $\hat{X}(\omega) = \delta(\omega - \omega_0)$, then on substituting $\hat{X}(\omega)$ in Equation (1.6) the following expression of $x(t)$ is obtained:

$$x(t) = \frac{1}{\sqrt{2\pi}}e^{-i\omega_0 t} \tag{1.9}$$

On substituting $x(t)$ on the right-hand side in Equation (1.5), the following equation is obtained, which is one of the definitions of the Dirac delta function.

$$\hat{X}(\omega) = \frac{1}{2\pi}\int_{-\infty}^{\infty} e^{i(\omega-\omega_0)t}dt = \delta(\omega-\omega_0) \tag{1.10}$$

The readers should also know about Parseval's theorem and Parseval's identity at this stage, as these will be used in later chapters. Parseval's theorem is written as follows:

$$\int_{-\infty}^{\infty} |x(t)|^2\, dt = \int_{-\infty}^{\infty} \left|\hat{X}(\omega)\right|^2 d\omega \tag{1.11}$$

This implies that the total energy content of a function $x(t)$ summed over all time t is equal to the total energy contained in its Fourier transform summed across all of its frequency components.

$$\frac{1}{\pi}\int_{-\pi}^{\pi} [f(t)]^2\, dt = \frac{a_0^2}{2} + \sum_{n=1}^{\infty} a_n^2 + b_n^2 \tag{1.12}$$

The identity tells us that the sum of the squares of the Fourier coefficients of a certain function, say $f(t)$, is equal to the integral of the square of the function.

1.3 RANDOM VIBRATION

The random vibration is a nondeterministic motion that has a unique randomness in its characteristic. The vibrations induced in trains and road vehicles due to track and road surface roughness, wind excitations, ground motions and wave loading are common examples of random vibrations. Typically, a random vibration may be either a stationary or a nonstationary process. A common characteristic feature of these vibrations is that these are randomly varying in time, which obviously means that these are nondeterministic (and hence nonperiodic) in nature. This implies that in case of random vibrations, it is not be possible to predict the amplitude of vibration accurately at any specific instant; however, one may predict the probability of occurrence of acceleration or displacement amplitude at an instant. Unlike a pure sinusoidal vibration, a random vibration contains a continuous spectrum of frequencies. It may be worth mentioning here that the histogram of a random datum or signal tends to cluster near its mean value, and the histogram itself takes the shape of a bell-shaped curve, which is very similar to the Gaussian distribution curve. The peak values in a random vibration data do not maintain any fixed ratios with the root mean square (RMS) values. Let us assume that a sample function is generated every time a test is performed. This sample function may be different every time the test is performed. This creates an ensemble of sample functions that constitutes what we refer to as a random or stochastic process. For example, during ground shaking a sensor at a particular place records some data (precisely the movement of ground with respect to time at a given interval). For a different earthquake at the same site, the same sensor measures different data and thus records a new accelerogram. In this way, for an infinite number of measurements, an infinite number of data sets would be generated. This is called an ensemble of accelerograms. Any random process may be characterized by some statistical parameters like mean value, standard deviation, kurtosis, etc. If these properties remain constant with time, then vibration may be said to be a random stationary process. However, if the estimates of these statistical parameters vary with time, the vibration then represents a nonstationary random process for which instantaneous value of the datum or signal can never be predicted at any points of time.

In case of a stationary stochastic (or random) process, say $N(t)$, the power spectral density function (PSDF) comes in very handy. The PSDF $S(\omega)$ is defined as the Fourier transform of the corresponding correlation function $R(\tau)$ and is written as

$$S(\omega) = \frac{1}{2\pi}\int_{-\infty}^{\infty} R(\tau)e^{i\omega\tau}d\tau \tag{1.13}$$

in which

$$R(\tau) = \frac{1}{T-\tau}\int_{0}^{T-\tau} N(\tau)N(t+\tau)dt \tag{1.14}$$

As $S(\omega)$ and $R(\tau)$ are even functions, their correlation function may be rewritten as

$$R(\tau) = \int_{-\infty}^{\infty} S(\omega)\cos(\omega\tau)d\omega \tag{1.15}$$

Putting $\tau = 0$ in Equation (1.15), one may write the following:

$$R(0) = E[N^2(t)] \int_{-\infty}^{\infty} S(\omega)d\omega$$

The above equation readily tells us that the total mean square of the random stationary process, $N(t)$, is obtained by finding the total area under the power spectral density curve.

If a body is subjected to an excitation force that is a random process, its response must also be a random process. Let us concentrate on a single degree of freedom system in which a certain mass m is connected to a spring (with stiffness coefficient k) and a dashpot (with coefficient, c) and the system is excited by an external time-varying force $f(t)$, which is an ensemble of several force-time histories. The system is shown in Figure 1.1, and the corresponding equation of motion is shown in Equation (1.16):

$$m\ddot{x}(t) + c\dot{x}(t) + kx(t) = f(t) \tag{1.16}$$

In the above equation, the dots represent differentiation with respect to time, and the terms m and $x(t)$ denote the mass and the displacement response of the system, respectively. Now, as the input excitation $f(t)$ has been assumed to be a random process containing uncertainties, we should obtain the response $x(t)$ also as a random process. Let us assume that this input excitation $f(t)$ to the system represents an earthquake base excitation. For our case, we assume this to be the same as one of the ground motions recorded during the Loma Prieta earthquake (1989) at the Dumbarton Bridge site, shown in Figure 1.2. It may be noted that this excitation has a mean value of 0.0539 mm/s^2, and the corresponding histogram is shown in Figure 1.3.

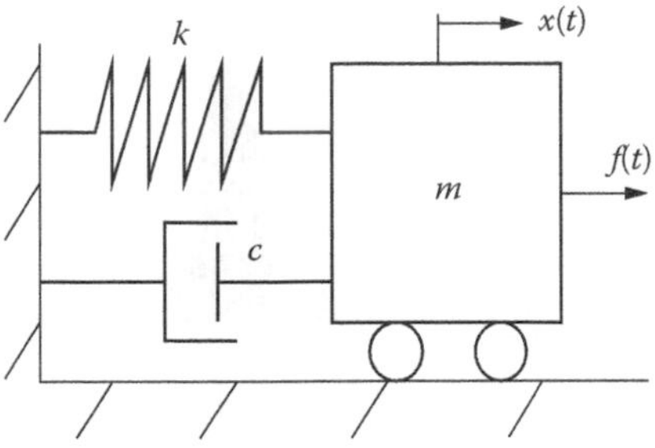

Figure 1.1 SDOF system subjected to external force.

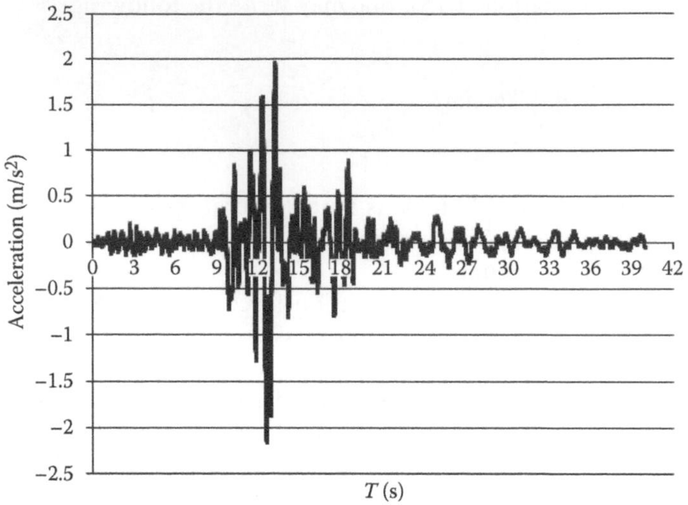

Figure 1.2 Acceleration time history record during the Loma Prieta earthquake (1989) near the Dumbarton Bridge site (Coyote Hills).

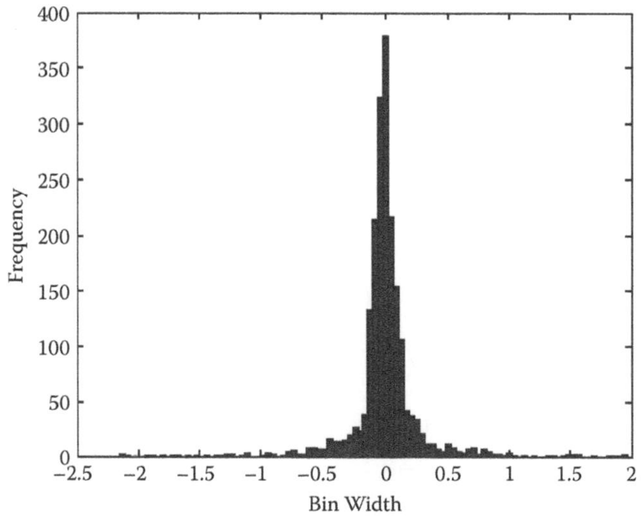

Figure 1.3 Histogram of the Loma Prieta earthquake motion.

1.4 WAVELET ANALYSIS

We have read in previous sections about Fourier transform and some important characteristics of random vibrations. The frequency spectrum gives in many cases thorough information about the sources of the signals that cannot be obtained from the time signal, e.g. the vibration level caused by rotating parts of a gearbox, the frequency content of vibration induced due to passage of high-speed trains underground, etc. Thus, a frequency analysis can be a valuable tool in locating the sources of vibrations as well as the level and frequency of vibrations. The Fourier transform, as discussed earlier, may work with periodic and nonperiodic functions, but these functions have to be stationary. The Fourier transform does not work with nonstationary functions or signals. However, it has become increasingly common in practice to analyse a nonstationary random signal or obtain responses of a structure when it is subjected to a nonstationary random excitation. The main purpose is to know the frequency content of a function at a certain time. The wavelet analytic tool has emerged as a powerful tool mainly due to its time–frequency localization property. Before going into the discussion on wavelets, it would be a good idea to understand the limitations of Fourier transform in analysis of signals that are nonstationary in nature.

Let us assume a stationary signal as shown in Figure 1.4. The signal has three frequency components, viz. 5, 10 and 25 Hz. Thus, this signal has at any instant of time the presence of all three frequencies. On performing Fourier transform of this signal, we may identify the existing frequencies

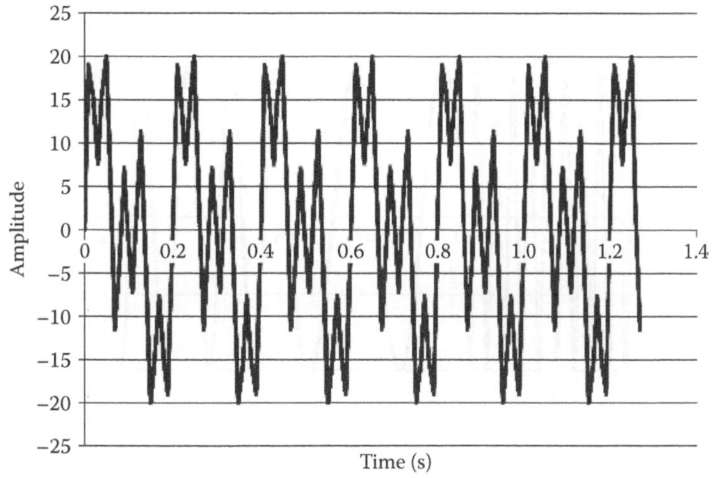

Figure 1.4 Stationary signal consisting of frequencies 5, 10 and 25 Hz.

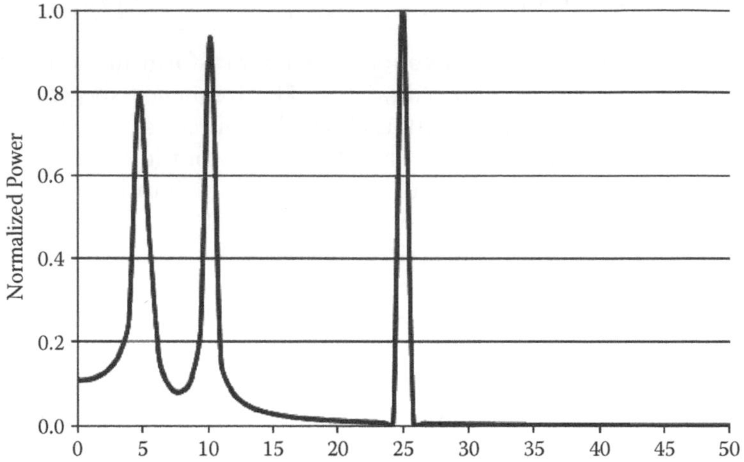

Figure 1.5 Frequency spectrum of the stationary signal with frequencies 5, 10 and 25 Hz.

as shown in Figure 1.5. Thus, Fourier transform is proved to be efficient in revealing the frequency content of a signal.

Now, let us assume a nonstationary signal as shown in Figure 1.6. This signal shows the variation of the frequency of its amplitude in three distinct time zones. From 0 to 0.4 s it has a frequency of 25 Hz, from 0.4 to 0.8 s the frequency is 5 Hz, and for the remaining duration the frequency is 10 Hz.

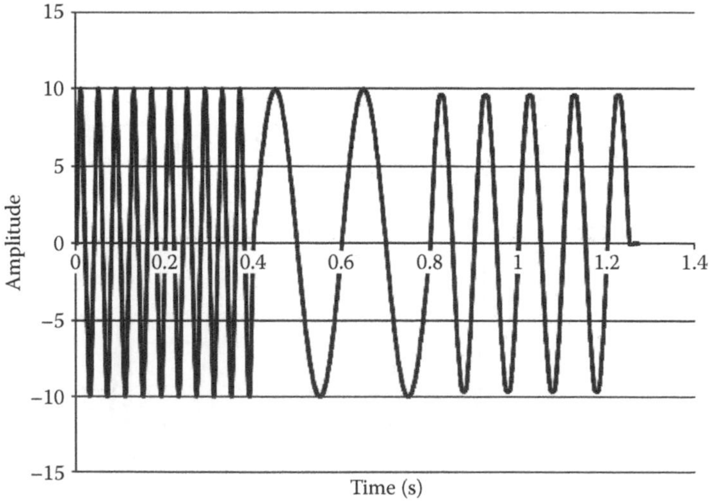

Figure 1.6 Nonstationary signal with frequencies 25, 5 and 10 Hz.

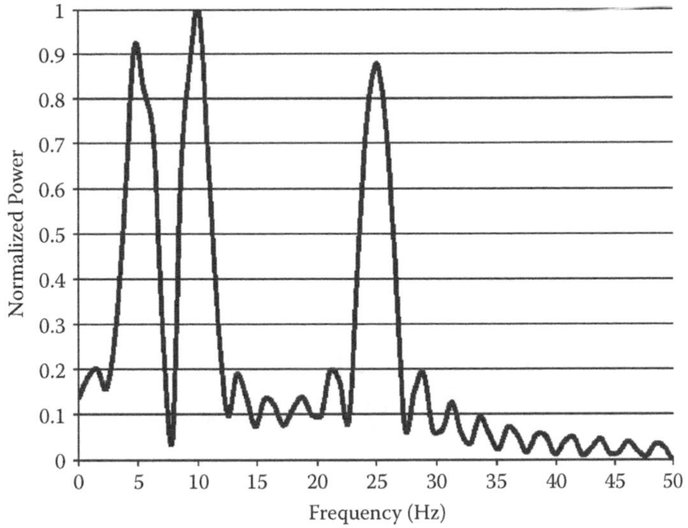

Figure 1.7 Frequency spectrum of the nonstationary signal with frequencies 25, 5 and 10 Hz.

These frequencies can be easily identified on taking the Fourier transform of the signal. The corresponding Fourier transform of the signal is shown in Figure 1.7. Let us also assume another nonstationary signal as shown in Figure 1.8. On performing Fourier transform of this signal, we can clearly see in Figure 1.9 the dominant peaks again at frequencies 5, 10 and 25 Hz.

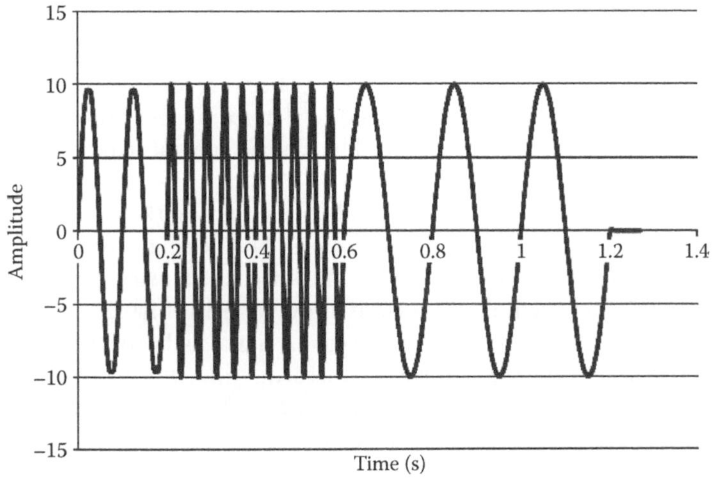

Figure 1.8 Nonstationary signal with frequencies 10, 25 and 5 Hz.

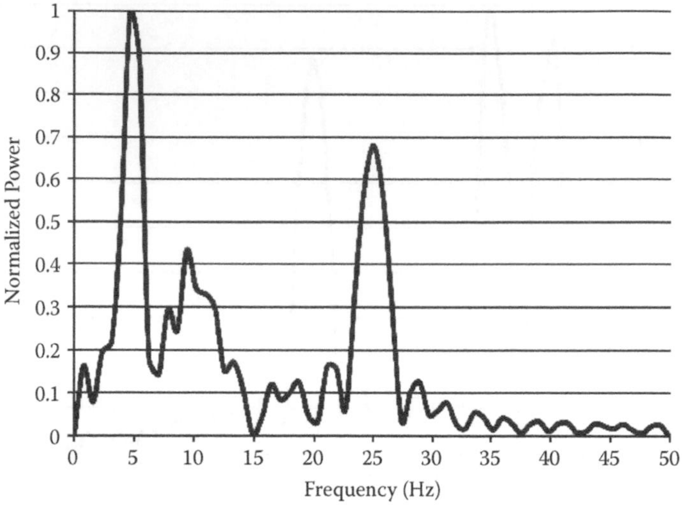

Figure 1.9 Frequency spectrum of the nonstationary signal with frequencies 10, 25 and 5 Hz.

Thus, we see that in all three signals, as shown in Figures 1.5, 1.7 and 1.9, the frequencies present are the same – 5, 10 and 25 Hz – but in different order. In the first case, the signal is stationary because the frequency content is the same at all instants of time. In the other two cases, though the frequency content is still the same, the occurrence of frequencies in the time domain is different. However, Fourier transformation of these nonstationary signals could not determine the order in which the frequencies appear in the time domain. This is a serious limitation of the Fourier transform in analysing stochastic nonstationary signals because in many physical and engineering problems we actually need to determine the presence of a frequency at a particular time instant. The Fourier transform clearly cannot handle this situation.

Thus, Fourier transform is not capable of extracting the information that we would like to know – time and frequency information – at the same time. Instead, what it does is generate the spectral density of the signal. It views the signal as a whole because it sums up all information over time; hence, the time information gets lost once a signal is Fourier transformed. Again, on taking inverse Fourier transform, one would integrate over the frequency range, and thus the frequency information is lost. So, the researchers tried to think in a more positive way and came up with an idea of short-time Fourier transform (STFT). This tool enabled the researchers to obtain the frequency content of a signal within a specified time interval. This could be achieved by breaking up the signal into several pieces and taking the Fourier transform of each piece. For this, a small window function, say $V(t)$, was necessary, which would help get the finite time variance and the finite frequency variance, respectively, as follows:

$$tV(t) \in \mathcal{L}_2(\mathbb{R}) \tag{1.17}$$

$$\frac{dV(t)}{dt} \in \mathcal{L}_2(\mathbb{R}) \tag{1.18}$$

In both Equations (1.17) and (1.18), $V(t) \in \mathcal{L}_2(\mathbb{R})$. So, the STFT of a function $x(t)$ with respect to the window $V(t)$ evaluated at translation τ_0 and modulation Ω_0 may be written as an inner or dot product as follows:

$$\text{STFT}(x(t), V(t)) = \int_{-\infty}^{\infty} x(t)\overline{V(t-\tau_0)}e^{-i\Omega_0 t}dt \tag{1.19}$$

which may further be rewritten as

$$\text{STFT}(x(t).V(t)) = \frac{1}{2\pi}\int_{-\infty}^{\infty} \hat{x}(\Omega)\overline{\hat{V}(\Omega-\Omega_0)}e^{j(\Omega-\Omega_0)\tau_0}d\Omega \tag{1.20}$$

The window may be any suitable function like a Gaussian function, raised cosine function, triangular window, etc. as shown in Figures 1.10 to 1.12. Thus, a properly translated window is actually used to multiply the function $x(t)$ to extract the information around $t = \tau_0$, and then the

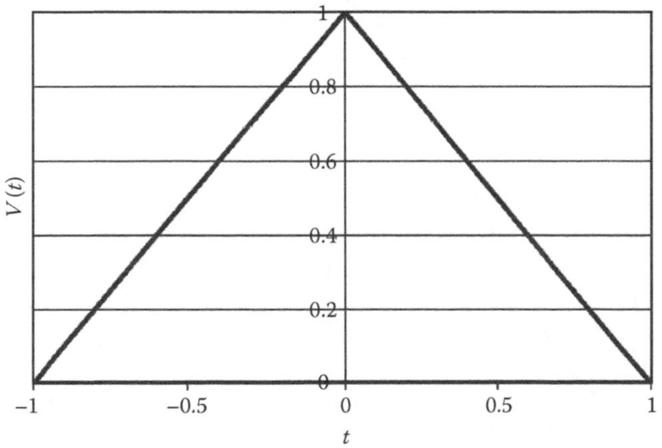

Figure 1.10 Triangular window.

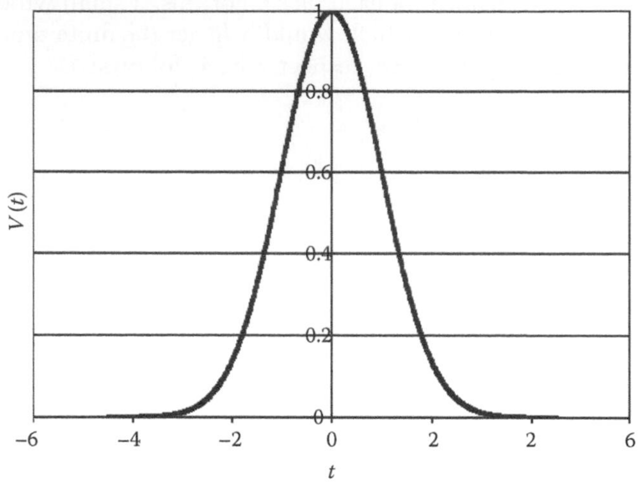

Figure 1.11 Gaussian window.

Fourier transform is taken; thus, it looks at the frequency content in a certain region. Equation (1.20) may actually be seen as Parseval's theorem, which says that the right-hand side of this equation is the dot product of the Fourier transforms of $x(t)$ and $\overline{V(t-\tau_0)e^{-i\Omega_0 t}}$. However, even this STFT concept was not good enough to capture time–frequency information together. One can easily notice the main shortcoming of the STFT from Figures 1.10 to 1.12, which is the limit in time and frequency. It

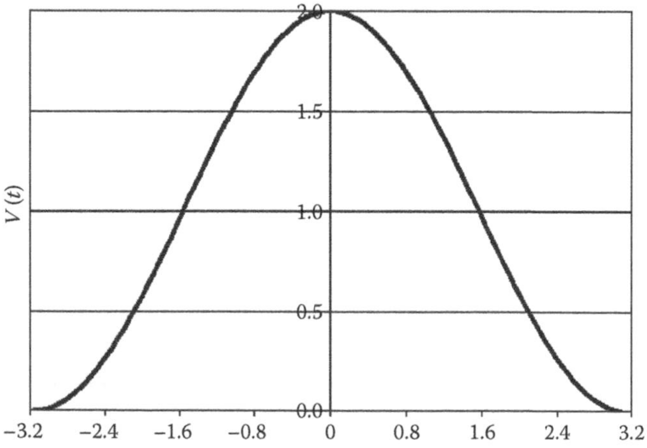

Figure 1.12 Raised cosine window.

means that one may indeed choose a suitable window, but cannot vary the size of the window in the time–frequency plane. So, when one multiplies the signal by a chosen window function (with finite length), it covers only a part of the signal, unlike in FT, where it is done over the whole time range. Thus, the frequency resolution becomes poorer. In fact, in STFT, narrow windows produce good time resolution but poor frequency resolution, and it is the opposite for wider windows. This is actually the reiteration of Heisenberg's uncertainty principle, which states that it is impossible to measure the position and momentum of a particle simultaneously with infinite precision. In the present context, this may interpreted as a fact that it is not possible to obtain good resolutions in both time and frequency simultaneously with a fixed window size. To overcome this limitation, one must use windows of different sizes, thus leading to multiresolution analysis.

At this point, the concept of wavelet transform comes to the rescue. The wavelet transform overcomes the resolution problem faced by STFT. The continuous wavelet transform (CWT) is the dot (inner) product of the signal $x(t)$ and a wavelet $\Psi(\frac{t-\tau}{s})$, where the wavelet (Ψ) is translated by the time parameter τ in the numerator and dilated by the frequency parameter s in the denominator. The term τ represents positive real numbers, and the term s represents both positive and negative real numbers. The continuous wavelet transform of a function $x(t)$ with respect to the mother wavelet Ψ evaluated at translation τ and scale s is defined as

$$\mathrm{CWT}_x(\tau,s) = \int_{-\infty}^{\infty} x(t)\frac{1}{\sqrt{s}}\left(\frac{t-\tau}{s}\right)dt \tag{1.21}$$

The mother wavelet, $(\frac{t-\tau}{s})$, in the above equation is a prototype from which the other window functions may be generated. All the wavelets are the dilated (scaling term is s, which is inverse of frequency) or compressed and shifted (translated in the time domain by an amount τ) versions of the mother wavelet. The most important characteristic feature of wavelet transform over STFT is that the width of the multiplying window is changed as the transform is computed for each and every spectral component. The window function in a wavelet transform is an oscillatory function and is compactly supported having finite length. The scaling term $\sqrt{s}$ in the denominator of Equation (1.21) is used for the purpose of energy normalization so that a wavelet-transformed signal will have the same energy at all scales. Let us have a closer look at this equation and try to understand how a wavelet transform actually works. Let us also assume that we have a signal $x(t)$ known to us. First, a specific scale is chosen and the wavelet is placed at the beginning of the signal when the time count starts ($t = 0$). The signal is then multiplied by the wavelet function and subsequently integrated over all times and divided by the factor $\sqrt{s}$ to get a single value called the wavelet coefficient for a specific scale and at a particular time point. The operation is repeated with the same scale, but now the wavelet function is placed at a different time instant, which is basically a shift from the previous position by a fixed amount, τ. This gives rise to the second wavelet coefficient. This goes on until one reaches the end of the signal. Thus, in doing so, one would ultimately obtain a row of wavelet coefficients for all time points but for a particular scale. Now, the scale is increased, and the whole process is repeated so as to obtain the second row of wavelet coefficients at all time points. In the end, when all desired values of scales are finished, one ends up with a complete array of wavelet coefficients for the signal $x(t)$. The product of the wavelet function with the signal at a particular location where the signal contains a spectral component generates larger amplitude; else the amplitude would be smaller or even zero.

Any function, say $y(t)$, may be expressed as a set of linear combinations of basis functions $\phi_k(t)$ with corresponding coefficients ν_k as follows:

$$y(t) = \sum_{j} \nu_k \phi_k(t) \tag{1.22}$$

Now, let us assume that $y(t)$ and $z(t)$ are two square integrable functions in the interval $[a, b]$ whose inner product is defined as

$$\langle y(t) z(t) \rangle \int_a^b y(t).\ z(t)dt \tag{1.23}$$

The CWT as shown in Equation (1.21) is the inner product of the signal $x(t)$ with the basis function $_{(\tau,s)}(t)$ where the $_{(\tau,s)}(t)$ is written as

$$_{(\tau,s)}(t) = \frac{1}{\sqrt{s}}\left(\frac{t-\tau}{s}\right) \tag{1.24}$$

The CWT depicts that the wavelet analysis is a measure of similar frequency content between the signal and its multiplier basis function. The wavelet coefficients generated from the calculation show the proximity of the signal to the wavelet basis function at the current scale. Thus, if there is a high correlation at a particular scale and time, the wavelet transform would generate a high value. When the wavelet is compressed, it represents high frequency.

There are several families of wavelets developed by researchers in the past and present. Out of all these wavelets, only three are discussed here: Mexican hat wavelet, Morlet wavelet and Daubechies wavelet (*db*2, as an example). The first two types may be used to perform continuous wavelet transform, while the third one is suitable for discrete wavelet transforms. The Mexican hat wavelet, Ψ_{mexh}, is the second derivative of the Gaussian probability density function as defined below:

$$_{mexh} = ce^{\left(-\frac{x^2}{2}\right)}(1-x^2) \tag{1.25}$$

where $c = \frac{2}{\sqrt{3}(\pi)^{0.25}}$. The Morlet wavelet, Ψ_{mor}, has the following expression:

$$_{mor} = e^{\left(-\frac{x^2}{2}\right)}\cos 5x \tag{1.26}$$

The Mexican hat wavelet and Morlet wavelet are shown in Figures 1.13 and 1.14, respectively.

When performing continuous wavelet transform of the nonstationary signals shown in Figure 1.6 (with frequencies in the order of 25, 5 and 10 Hz) and Figure 1.8 (with frequencies in the order of 10, 25 and 5 Hz) using Daubechies wavelet *db*2, one may obtain the two graphs (three-dimensional plots) shown in Figures 1.15 and 1.16, which show the variation of wavelet coefficients with time.

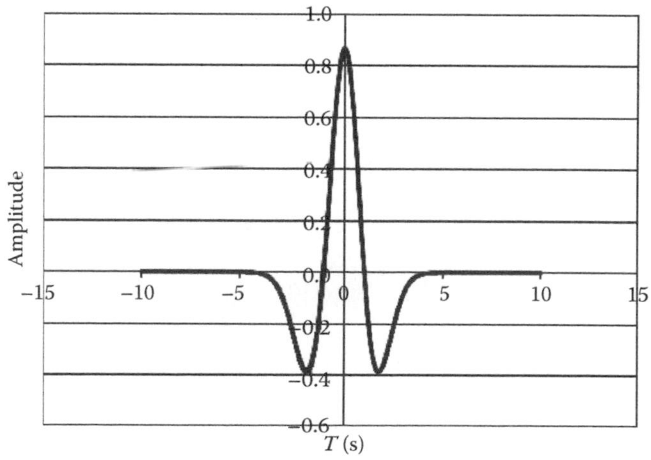

Figure 1.13 Mexican hat wavelet.

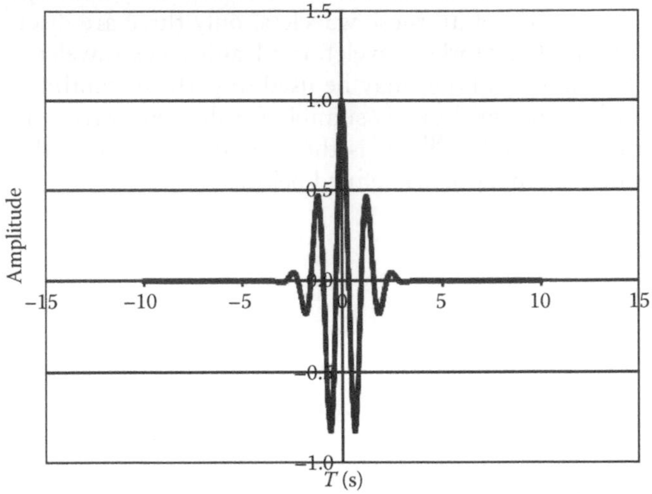

Figure 1.14 Morlet wavelet.

The two plots of wavelet coefficients of signals shown in Figures 1.15 and 1.16 clearly show the occurrence of frequencies at corresponding time regions. So, one can easily identify which frequencies are showing up at what frequencies. This is the most important feature of wavelet-based analysis. This unique time–frequency localization property has made the wavelets unique and most widely used for various purposes in the fields of signal processing, speech recognition, compression-decompression, image

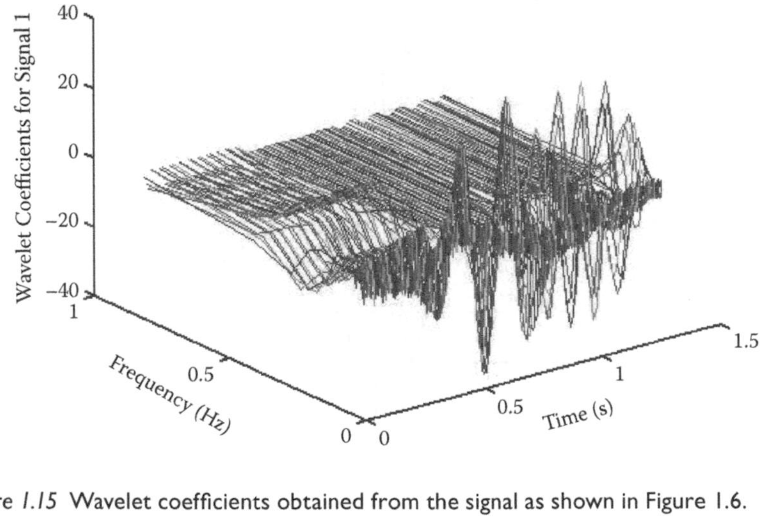

Figure 1.15 Wavelet coefficients obtained from the signal as shown in Figure 1.6.

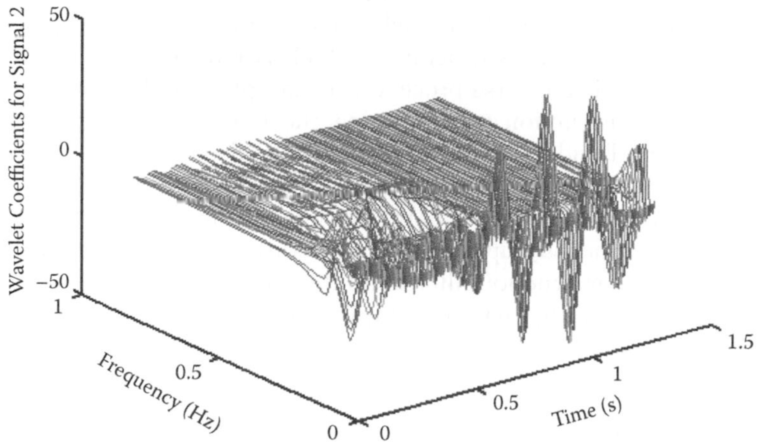

Figure 1.16 Wavelet coefficients obtained from the signal as shown in Figure 1.8.

processing, etc. In the next chapter we shall see how wavelets may be used to characterize ground motions and how the response of a single-degree-of-freedom system could be obtained performing the analysis in the wavelet domain.

1.5 A BRIEF REVIEW OF WAVELET PROPERTIES

In this section we are going to discuss some important properties of wavelet transform that will be used later in this book. The wavelet transform uses irregular shaped signals as basis functions that help in capturing discontinuities in the signal and abrupt changes. If a function $f(t)$ belongs to a space of all finite energy functions satisfying the following condition, the function can be decomposed by wavelet coefficients using wavelet transformation and later may be reconstructed from these coefficients using inverse wavelet transformation.

$$\int_{-\infty}^{\infty} |f(t)|^2 \, dt < \infty \tag{1.27}$$

The choice of a wavelet function may be linked to the characteristics and properties of the signal. The discrete wavelet scheme is used for wavelet decomposition of a signal and its reconstruction. During this process, a signal is decomposed in high-scale (low-frequency) components called trends and low-scale (high-frequency) components called details. This decomposition is based on low- and high-pass filters, depending on the wavelet basis function used. At each stage (scale) of filtering, twice the amount of original data is generated, which is reduced by a factor of 2 (downsampling). The inverse process of upsampling and filtering is used during the reconstruction process using the low- and high-pass reconstruction filters. Besides wavelet function, there also exists a scaling function that is associated with the wavelet analysis of a signal. This scaling function is related to the trends of the wavelet decomposition. The scaling function is obtained by upsampling and convolving the low-pass with the high-pass reconstruction filter in an iterative manner. Daubechies [8] has given the following dilation equations for scaling ($\varphi(t)$) and wavelet functions ((t)).

$$\varphi(t) = \sum c_k \varphi(2t - k) \tag{1.28}$$

$$(t) = \sum (-1)^k c_{1-k} \varphi(2t - k) \tag{1.29}$$

The uniqueness is defined as $\Sigma c_k = 2$. The details about wavelet-based decomposition and reconstruction of signals may be found in the work by Grossman and Morlet [4].

1.5.1 Continuous and discrete wavelet transform

The short-time Fourier transform (STFT) of a function $f(t)$ may be represented by the following equation, which is another form of Equation (1.19):

$$F(\omega, t_0) = \int_{-\infty}^{\infty} f(t) e^{-i\omega t} g(t - t_0) dt \tag{1.30}$$

Using this STFT definition, the wavelet function may be defined in the following manner. Let it be assumed that $g(t)$ is a window function that depends on time and frequency, and hence is defined as follows:

$$g(t) = |\omega|^{\frac{1}{2}} g(\omega t) \tag{1.31}$$

On substituting Equation (1.31) in Equation (1.30), one may get

$$\tilde{F}(\omega, t_0) = |\omega|^{\frac{1}{2}} \int_{-\infty}^{\infty} f(t) e^{-i\omega t} g(\omega(t - t_0)) dt \tag{1.32}$$

On dividing Equation (1.32) by $e^{-i\omega t_0}$ and later replacing ω with $\frac{1}{a}$ (thus, scale becomes inverse of frequency), one may obtain the following:

$$\frac{\tilde{F}(\omega,t_0)}{e^{-it_0/a}} = \frac{1}{|\alpha|^{\frac{1}{2}}} \int_{-\infty}^{\infty} f(t)e^{-i(t-t_0)/a} g\left(\frac{t-t_0}{a}\right)dt \tag{1.33}$$

From the last equation, the wavelet function may be identified as $(t) = e^{it} g(t)$, which represents the mother wavelet. On using translating and dilating parameters (*b* and *a*, respectively) in the mother wavelet, a series of other wavelets (of the same family) may be obtained. The term *b* precisely denotes the time position around which the wavelet basis function is centrally placed, and the term *a* represents the decomposition level (scale) of the wavelet transform. The Fourier transform of the mother wavelet basis function is given by

$$_{\hat{a},b}(\omega,t_0) = e^{ib\omega} \frac{1}{\hat{a}}(\omega) \tag{1.34}$$

With these definitions, Equation (1.33) takes the following form, which is the definition of the continuous wavelet transform [4]:

$$_{\hat{a},b}(\omega,t_0) = \frac{1}{|a|^{1/2}} \int_{-\infty}^{\infty} f(t)\left(\frac{t-t_0}{a}\right)dt \tag{1.35}$$

For the wavelet transform to contain complete information of $\ddot{x}_g(t)$, the normalization constant has to be of finite value ($C_\psi < \infty$), which obviously means that $\int_{-\infty}^{\infty} \psi_{a,b}(t)dt = 0$. This, in turn, emphasizes that the wavelet function has finite energy and hence must vanish. This is called the admissibility condition of wavelet transform.

The discrete wavelet transform helps in analysing discrete signals or sampling continuous signals. The scales and time locations of the wavelet functions in case of discrete wavelet transform cannot take arbitrary values because the discretization rule obeyed by the specific wavelet scheme specifies the scales and time positions. The wavelet functions at various scales have variable size (but constant shape). The wavelet family in such a case is defined as

$$_{j,k}(t) = a_0^{j/2}(a_0^j(t-b)) \tag{1.36}$$

The dyadic wavelet scheme is another useful discretization scheme. In this case, the scale parameter is sampled on a dyadic grid. The corresponding discrete wavelet transform is defined as follows:

$$_{\hat{a},b}(a_j,b_i)=2^{j/2}\int_{-\infty}^{\infty}f(t)(2^j t-k)dt \tag{1.37}$$

In the above equation for discrete wavelet transform, $a_j = 2^j$ and $b_i = k2^j$. The integral in Equation (1.37) is performed in discrete format using the pyramid algorithm or fast wavelet algorithm [5, 6]. Usually, discrete wavelet transform (DWT) provides an efficient representation of a stochastic function compared to the representation by continuous wavelet transform (CWT) because DWT uses fewer scales and time positions to decompose a signal. The CWT helps in identification of singularities in the signals that might be ignored by a discrete grid. The computation of integrals for CWT and DWT is, however, not discussed here.

1.5.2 Vanishing moments

Another important property of the wavelet transform is the number of vanishing moments that represent the regularity of the wavelet functions and ability of a wavelet transform to capture localized information. A wavelet $\Psi(t)$ has M vanishing moments if the following condition is satisfied:

$$\int_{-\infty}^{\infty}t^m(t)dt=0;\quad m=0,1,2,\ldots,M-1 \tag{1.38}$$

The number of vanishing moments is directly related to the regularity of the wavelet. A more regular wavelet has a greater number of vanishing moments. The regularity of a function is expressed by the order of differentiability. For most of the wavelet functions, such as the Daubechies family of wavelets, the number of vanishing moments and the degree of differentiability are known. A wavelet of zero mean has at least one vanishing moment. A less regular wavelet function is suited more for the analysis of nonstationary data, whereas a more regular or smoother wavelet function is suitable for the analysis of stationary data. The localization property is another important feature of wavelets that helps capture the localized effects in the time domain as well as the frequency domain. Localization and regularity are inversely related to each other. Thus, depending on the specific characters of a signal to be analysed, preference between regularity and localization has to be made.

1.5.3 Uncertainty principle

The resolution in the time–frequency plane is limited by the uncertainty principle. The uncertainty principle states that contraction in the time domain dilates the function in the frequency domain. This means any effort to improve time localization will come at the cost of poor frequency resolution. This is as per the Heisenberg uncertainty principle, which is expressed as

$$\sigma_t \sigma_p \geq \frac{1}{2} \tag{1.39}$$

This signifies that a reduction in the value of the standard deviation in time (thereby improving time resolution) must increase the value of the standard deviation in frequency (thus reducing frequency resolution). Thus, the user has to decide the priority of one over the other and select the wavelet functions accordingly. The Shannon wavelet provides a good frequency resolution and is defined by

$$(t) = \frac{\sin\{2\pi(t-0.5)\}}{2\pi(t-0.5)} - \frac{\sin\{\pi\}}{\pi(t-0.5)} \tag{1.40}$$

The Shannon wavelet function is infinitely differentiable, and it has an infinite number of vanishing moments. The Fourier transform of the Shannon wavelet results in the following:

$$(\hat{\omega}) = e^{-i\frac{\omega}{2}}; \quad \omega \in [-2\pi, -\pi] \cup [\pi, 2\pi] \tag{1.41}$$

However, Shannon wavelets have poorer time resolution. Daubechies wavelets also offer good frequency resolution with increasing order. On the other hand, harmonic wavelets may give good time and frequency resolution together (to some extent), depending on suitable values of the parameters of harmonic wavelets. An example of a harmonic wavelet and its Fourier transform is given below [9]:

$$f(t) = \int_{-\infty}^{\infty} \hat{F}(\omega) e^{i\omega t} d\omega = \frac{e^{i4\pi\omega t} - e^{i2\pi\omega t}}{i2\pi t} \tag{1.42}$$

On taking Fourier transform, the frequency domain representation of the harmonic wavelet function may be written as

$$\hat{F}(\omega) = \frac{1}{2\pi}; \quad 2\pi \leq \omega \leq 4\pi \qquad (1.43)$$
$$= 0; \quad \text{otherwise}$$

Though the harmonic wavelet is a complex function with real and imaginary components as even function and odd function, respectively, it has practically a finite support (hence, it contains finite energy).

Several authors in the last few decades have written quite a good number of books on wavelets [9–11]. However, a book on application of wavelets, especially with reference to civil engineering problems, is not really known to exist. This book may be viewed as a sincere effort to fill in that gap.

Chapter 2

Vibration analysis of SDOF and MDOF systems in the wavelet domain

The simplest approach in studying the response of a structural system to an external dynamic excitation is to idealize the system as a linear single-degree-of-freedom (SDOF) system comprising a mass body, a spring representing the system stiffness and a dashpot representing the energy dissipation mechanism of the system. In many of the cases, the dynamic excitation to which the system is subjected is a seismic ground motion process. The seismic motions are known to be highly nonstationary with time-varying statistical characteristics and are also characterized by time-dependent frequency content due to the dispersion of the constituent waves. Different solution procedures have been proposed to obtain the nonstationary system response; however, the specific modulating functions used in the approach determined the nature of the nonstationarity to be considered in the analysis. Moreover, the time-varying frequency content of input excitations has not been considered. With the development of the wavelet-based analytical technique, it has become possible to tackle the frequency nonstationarities as well. In this chapter, we will see how to use the discretized version of the continuous wavelet transform to obtain the nonstationary response of a SDOF system subjected to seismic ground motion excitation. The basic idea to solve system equations of motion in the wavelet domain is clearly explained in this chapter.

2.1 WAVELET-BASED DISCRETIZATION OF GROUND MOTIONS

The seismic ground motion processes are transient and contain finite energy. Thus, these processes may be well represented by statistical functionals of wavelet coefficients. Let us consider the earthquake excitation

$\ddot{x}_g(t)$ as a zero mean process with nonstationary Gaussian characteristics. The wavelet transform of this process is written as [8]

$$W_\psi \ddot{x}_g(a,b) = \frac{1}{\sqrt{|a|}} \int_{-\infty}^{\infty} \ddot{x}_g(t) \psi^* \left(\frac{t-b}{a} \right) dt, \quad a > 0 \tag{2.1}$$

The inverse wavelet transform would fetch the ground motion process $\ddot{x}_g(t)$ as

$$\ddot{x}_g(t) = \frac{1}{2\pi C_\psi} \int_{-\infty}^{\infty} \int_{-\infty}^{\infty} \frac{1}{a^2} W_\psi \ddot{x}_g(a,b) \psi_{a,b}(t)\, da\, db \tag{2.2}$$

The asterisk in Equation (2.1) denotes the complex conjugation. In Equation (2.2), the term C_ψ must satisfy the following admissibility condition in the frequency domain:

$$C_\psi = \int_{-\infty}^{\infty} \frac{|\hat{\psi}(\omega)|^2}{|\omega|} d\omega < +\infty \tag{2.3}$$

In the above equations, the function ψ is the basic or mother wavelet. $\hat{\psi}$ indicates the Fourier transform of ψ. This condition implies that the wavelet integrates to zero, and this requires that the wavelet must have m vanishing moments, i.e.

$$\int_{-\infty}^{\infty} t^k \psi(t) dt = 0; \quad k = 0, \ldots, m \tag{2.4}$$

During a certain time a wavelet oscillates like a wave and is then localized. The oscillation of a wavelet is determined by the number of vanishing moments. The localization of the wavelet is measured by the interval where it has values other than zero. From this single function ψ a family of functions may be constructed using translation and dilation. The term $\hat{\psi}$ is written as

$$\hat{\psi}(\omega) = \frac{1}{\sqrt{2\pi}} \int_{-\infty}^{\infty} \psi(t) e^{-i\omega t} dt \tag{2.5}$$

The wavelet functions $\psi_{a,b}(t)$ are constructed by the translated and dilated version of the mother wavelet using parameters a and b as

$$\psi_{a,b}(t) = \frac{1}{\sqrt{|a|}}\left(\frac{t-b}{a}\right); \quad a,b \in R^{+} \tag{2.6}$$

The translation parameter b centres the basis function at $t = b$ by windowing over a certain temporal stretch around b depending on the parameter a. The frequency content of the wavelet basis function $\psi_{a,b}(t)$, obtained by its Fourier transform, can be controlled by the parameter a. Since the basis functions convolute with the function $\ddot{x}_g(t)$ in Equation (2.1), the information about the frequency content of the said function at the point where the basis function is centred is contained in the corresponding wavelet coefficients [12]. Thus, the wavelet transform coefficients $W_\psi \ddot{x}_g(a,b)$ of the ground motion process represent the contribution to the process in the neighbourhood of $t = b$ and in the frequency band corresponding to the dilation factor of a. The integrals used in Equations (2.1) and (2.2) must be discretized for the purpose of numerical computation. For this purpose, the following assumptions are necessary:

$$a_j = \sigma^j \tag{2.7}$$

$$b_j = (j-1)\Delta b \tag{2.8}$$

By adopting a discretization scheme [13] and using the above assumptions, the following expressions are obtained:

$$\Delta a_j = 0.5 * (a_{j+1} - a_j) + (a_j - a_{j-1}) = \frac{a_j}{2}\left(\sigma - \frac{1}{\sigma}\right) \tag{2.9}$$

$$\Delta b_j = 0.5 * (b_{j+1} - b_j) + (b_j - b_{j-1}) = \Delta b \tag{2.10}$$

Using the expressions for Δa_j and Δb, the ground motion process as shown in Equation (2.2) may be rewritten in a discretized version as

$$\ddot{x}_g(t) = \sum_i \sum_j \frac{K\Delta b}{a_j} W_\psi \ddot{x}_g(a_j, b_i)\psi_{a_j,b_i}(t) \tag{2.11}$$

In the above equation, the term K has the following expression:

$$K = \frac{1}{2\pi C_\psi}\left(\sigma - \frac{1}{\sigma}\right) \tag{2.12}$$

Equation (2.11) is particularly important because it shows that any nonstationary stochastic process may be discretized in terms of wavelet coefficients of the process. This concept will be used in this and the following chapters of the book. Before going into the formulation of the SDOF problem and its solution in the wavelet domain, it is worth discussing at this stage the time–frequency characteristics of wavelets that will be used quite often in almost all chapters.

2.2 TIME–FREQUENCY CHARACTERISTICS OF WAVELETS

The time-varying second-order moment statistics of the ground motion process $\ddot{x}_g(t)$ are obtained on taking its inner product [8], which may otherwise be written as follows:

$$\int_{-\infty}^{\infty} \ddot{x}_g^2(t)dt = \frac{1}{2\pi C_\psi} \int_{-\infty}^{\infty}\int_{-\infty}^{\infty} \frac{1}{a^2}[W_\psi \ddot{x}_g(a,b)^2]\, da\, db \tag{2.13}$$

The double integral in Equation (2.13) may be discretized using the discretization scheme as shown by Equations (2.9) and (2.10), and then on taking expectations of both sides, the following equation is obtained:

$$E\left[\ddot{x}_g^2(t)\right] = K\sum_i \sum_j \frac{\Delta b}{a_j} E[W_\psi \ddot{x}_g(a_j,b_i)^2] \tag{2.14}$$

It should be noted here that the wavelet coefficients $W_\psi \ddot{x}_g(a_j,b_i)$ give localized information of the process $\ddot{x}_g$ about time instant $t = b_i$. So, from Equation (2.14), the instantaneous mean square value is derived as

$$E\left[\ddot{x}_g^2(t)\right]_{t=b_i} = K\sum_i \frac{E[W_\psi \ddot{x}_g(a_j,b_i)^2]}{a_j} \tag{2.15}$$

The expected integral growth $\xi(t)\,|_{t=b_i}$ in the temporal energy up to the time instant $t = b_i$ is obtainable from the integral mean square response as shown below:

$$\xi(t)|t = b_i = \int_0^{t=b_i} \ddot{x}_g^2(\tau)d\tau = K\sum_j \sum_{i=0}^{i} \frac{E[W_\psi \ddot{x}_g(a_j,b_i)^2]}{a_j} \tag{2.16}$$

The mean square integral response up to the time instant $t = b_i$ corresponding to a specific frequency band with $a_j = \sigma^j$ is obtainable as well from the following equation:

$$\xi(t)\Big|_{t=b_i}^{j} = K\sum_{i=0}^{i} \frac{E[W_\psi \ddot{x}_g(a_j,b_i)^2]}{a_j} \tag{2.17}$$

At this stage we are interested in finding out the total expected energy of the ground motion process. This is obtained by using the Parseval identity in Equation (2.14) in conjunction with the following orthogonal relation:

$$\int_{-\infty}^{\infty} \hat{\psi}_{a_j,b_i}(\omega)\hat{\psi}^{*}_{a_j,b_l}(\omega)d\omega = \delta_{jk}\delta_{il} \tag{2.18}$$

The orthogonal relation will be explained later in the chapter. The total expected energy of the process thus will take the following expression:

$$\int_{-\infty}^{\infty} E[|\ddot{X}(\omega)|^2]d\omega = \int_{-\infty}^{\infty} \sum_i \sum_j \frac{K\Delta b}{a_j} E\left[W_\psi \ddot{x}_g(a_j,b_i)^2\right]\left|\hat{\psi}_{a_j,b_l}(\omega)\right|^2 d\omega \tag{2.19}$$

In the above equation, $\ddot{X}(\omega)$ denotes the Fourier transform of the ground motion process, $\ddot{x}_g(t)$. The expectation of the squared wavelet coefficients $E[W_\psi \ddot{x}_g(a_j,b_i)^2]$ correspond to nonoverlapping energy bands for different j values. Thus, the expected energy of $\ddot{x}_g(t)$ in the frequency band corresponding to a_j is given as

$$\int_{-\infty}^{\infty} E[|\ddot{X}^j(\omega)|^2]\, d\omega = \int_{-\infty}^{\infty} \sum_i \frac{K\Delta b}{a_j} E[W_\psi \ddot{x}_g(a_j,b_i)^2]\, |\hat{\psi}_{a_j,b_l}(\omega)|^2\, d\omega \tag{2.20}$$

The above equation may also be rewritten as

$$\int_{-\infty}^{\infty} E[|\ddot{X}^j(\omega)|^2]d\omega = \sum_i \frac{K\Delta b}{a_j} E[W_\psi \ddot{x}_g(a_j,b_i)^2] \tag{2.21}$$

so that

$$\int_{-\infty}^{\infty} E[|X(\omega)|^2]d\omega = \sum_j \int_{-\infty}^{\infty} E[|F^j(\omega)|^2]d\omega \tag{2.22}$$

2.3 FORMULATION OF SDOF SYSTEM EQUATION IN WAVELET DOMAIN

Let us imagine a single-storey building structure that may be represented by a SDOF system as shown in Figure 2.1.

The SDOF system contains the lumped mass m_s of the building at a certain height. The stiffness of the columns supporting the building and the viscous damping coefficient are represented by the terms k_s and c_s, respectively. It may be mentioned here that the stiffness and the damping are dependent on the mass and the natural frequency of vibration, ω_n, of the structure along with its damping ratio, ζ. The relevant expressions are given below:

$$k_s = m_s \omega_n^2 \tag{2.23}$$

$$c_s = 2\zeta_s m_s \omega_n \tag{2.24}$$

Let the building be subjected to a horizontal earthquake excitation process $\ddot{x}_g(t)$ at the base. The equation of motion for this system is written as

$$m_s \ddot{x}_s(t) + c_s \dot{x}_s(t) + k_s x_s(t) = -m_s \ddot{x}_g(t) \tag{2.25}$$

The terms $x_s(t)$, $\dot{x}_s(t)$ and $\ddot{x}_s(t)$ denote the relative displacement, relative velocity and relative acceleration of the structure in the horizontal direction with respect to the ground. On substituting the expressions for k and c from Equations (2.23) and (2.24) into Equation (2.25), and eliminating the mass term from both sides, the simplified equation takes the following form:

$$\ddot{x}_s(t) + 2\zeta_s \omega_n \dot{x}_s(t) + \omega_n^2 x_s(t) = -\ddot{x}_g(t) \tag{2.26}$$

The next goal would be to solve this dynamic equation of the system in the wavelet domain. This requires some additional tasks to be done. The

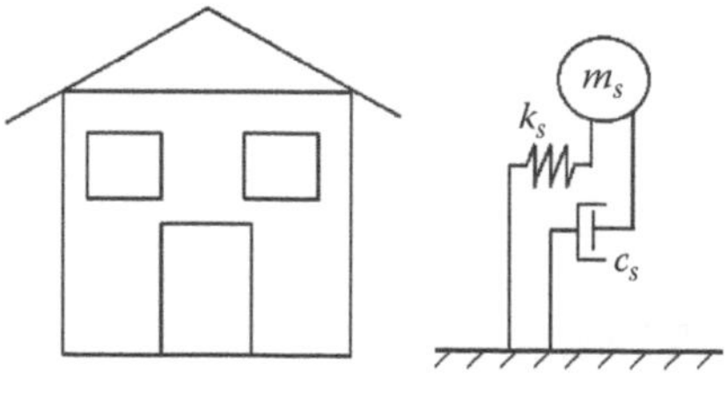

Figure 2.1 Single-storey building idealized as SDOF system.

first and foremost is to expand the structural lateral displacement response term, $x_s(t)$, in terms of wavelet functions. Similar to Equation (2.11), which expresses the ground acceleration in terms of wavelet coefficients, the lateral displacement of the structure may also be expressed in terms of wavelet functionals as follows:

$$x_s(t) = \sum_i \sum_j \frac{K\Delta b}{a_j} W_\psi x_s(a_j, b_i) \psi_{a_j,b_i}(t) \tag{2.27}$$

On differentiating the above equation one and two times with respect to time, the velocity and acceleration terms, $\dot{x}_s(t)$ and $\ddot{x}_s(t)$, may be respectively obtained as

$$\dot{x}_s(t) = \sum_i \sum_j \frac{K\Delta b}{a_j} W_\psi x_s(a_j, b_i) \dot{\psi}_{a_j,b_i}(t) \tag{2.28}$$

$$\ddot{x}_s(t) = \sum_i \sum_j \frac{K\Delta b}{a_j} W_\psi x_s(a_j, b_i) \ddot{\psi}_{a_j,b_i}(t) \tag{2.29}$$

The relative displacement, velocity and acceleration terms in Equation (2.26) may be replaced with corresponding wavelet coefficients as shown in Equations (2.27) to (2.29) to get the following transformed equation:

$$\begin{aligned} &\sum_i \sum_j \frac{K\Delta b}{a_j} W_\psi x_s(a_j, b_i) \ddot{\psi}_{a_j,b_i}(t) + 2\zeta_s \omega_n \sum_i \sum_j \frac{K\Delta b}{a_j} W_\psi x_s(a_j, b_i) \ddot{\psi}_{a_j,b_i}(t) \\ &\quad + \omega_n^2 \sum_i \sum_j \frac{K\Delta b}{a_j} W_\psi x_s(a_j, b_i) \psi_{a_j,b_i}(t) = -\sum_i \sum_j \frac{K\Delta b}{a_j} W_\psi \ddot{x}_g(a_j, b_i) \psi_{a_j,b_i}(t) \end{aligned} \tag{2.30}$$

On cancelling $K\Delta b$ and rearranging the terms, the above equation may be simplified to

$$\begin{aligned} &\sum_i \sum_j \frac{1}{a_j} W_\psi x_s(a_j, b_i) \left\{ \ddot{\psi}_{a_j,b_i}(t) + 2\zeta_s \omega_n \dot{\psi}_{a_j,b_i}(t) + \omega_n^2 \psi_{a_j,b_i}(t) \right\} \\ &\quad = -\sum_i \sum_j \frac{1}{a_j} W_\psi \ddot{x}_g(a_j, b_i) \psi_{a_j,b_i}(t) \end{aligned} \tag{2.31}$$

Our next objective would be to transform the above equation from the time domain to the frequency domain, and hence we perform Fourier transform to get the following equation:

$$\sum_i \sum_j \frac{1}{a_j} W_\psi x_s(a_j, b_i)\left\{-\omega^2 \hat{\psi}_{a_j,b_i}(\omega) + 2i\zeta_s \omega\omega_n \hat{\psi}_{a_j,b_i}(\omega) + \omega_n^2 \hat{\psi}_{a_j,b_i}(\omega)\right\}$$

$$= -\sum_i \sum_j \frac{1}{a_j} W_\psi \ddot{x}_g(a_j, b_i)\hat{\psi}_{a_j,b_i}(\omega) \tag{2.32}$$

The above equation may be simplified further to obtain the lateral relative displacement of the SDOF system considered in terms of wavelet coefficients as follows:

$$\sum_i \sum_j \frac{1}{a_j} W_\psi x_s(a_j, b_i)\hat{\psi}_{a_j,b_i}(\omega) = \sum_i \sum_j \frac{1}{a_j} W_\psi \ddot{x}_g(a_j, b_i) H(\omega)\hat{\psi}_{a_j,b_i}(\omega) \tag{2.33}$$

In the above equation, $H(\omega)$ is the frequency-dependent transfer function relating the relative horizontal displacement, x_s, of the structure to the ground acceleration process in terms of wavelet coefficients and is written below:

$$H(\omega) = -\frac{1}{(\omega_n^2 - \omega^2) + i(2i\zeta_s \omega\omega_n)} \tag{2.34}$$

2.4 WAVELET BASIS FUNCTION FOR GROUND MOTION PROCESS

We are solving a problem where a SDOF system is subjected to a nonstationary random seismic motion. The solution in the wavelet domain depends on the wavelet coefficients of this ground motion process. Therefore, we must use a wavelet basis function that would be suitable for this application. The choice of wavelet depends upon this application and the characteristics of the function that would be analysed. For the current problem, we will use an orthogonal wavelet basis function as proposed by Basu and Gupta [12]. This function is a modified form of the Littlewood–Paley (L-P) basis. The L-P basis, though, does not have good time-localization ability, but has a very good frequency localization property. The wavelet transform is characterized by its Fourier transform as [12]

$$\hat{\psi}(\omega) = \begin{cases} \dfrac{1}{\sqrt{2(p-1)\pi}}, & \pi \le |\omega| \le p\pi \\ 0, & \text{otherwise} \end{cases} \tag{2.35}$$

For $p = 2$, the above function becomes the same as the L-P basis. However, with this value of p, $\hat{\psi}(\omega)$ remains constant over a wide frequency band. In that case, the L-P basis cannot be used to represent realistic ground motions. The value of p should be chosen wisely to correctly represent the fluctuations in the frequency spectrum in case of seismic motion processes. After conducting detailed investigations on the probable value of p, the value of $2^{1/4}$ is proposed by Basu and Gupta [12], as this value has been found to be reasonable for most of the ground motions. Any smaller value than this would lead to increased computation effort due to the presence of a greater number of energy bands. However, a value of p greater than the proposed value of $2^{0.25}$ may be chosen as well to reduce the computational burden in case the ground motions have relatively smooth Fourier spectra. It must be mentioned here that the value of σ must be kept the same as the value of p (i.e. $\sigma = p$); otherwise, the functions generated by the dilated and translated versions of the wavelet function may not be mutually orthogonal, which is a desirable feature for a wavelet basis function without which the analysis of the system becomes complicated.

On performing inverse Fourier transform of Equation (2.35) and using the relation $\sigma = p$, the proposed wavelet basis function becomes

$$\psi(t) = \frac{1}{\pi\sqrt{\sigma - 1}} \frac{\sin(\sigma\pi t) - \sin(\pi t)}{t} \tag{2.36}$$

The orthogonality relation for the proposed modified L-P basis is given as

$$\int_{-\infty}^{\infty} \hat{\psi}_{a_j,b_i}(\omega)\hat{\psi}^{*}_{a_j,b_i}(\omega) = \delta_{jk}\delta_{il} \tag{2.37}$$

The term δ_{mm} denotes the Kronecker delta function, which is 1 when the two variables are the same or 0 when they are different. The advantage of the modified L-P basis is that it is more useful in capturing energy contribution to each frequency band. It also facilitates representation of input and output instantaneous power spectral density functions (PSDFs) due to the nonoverlapping frequency bands for different values of dilation parameters. This also ensures that due to the input from a specific frequency band, only this particular band is excited in the output, and not any other band, and this makes the point-wise calculation of instantaneous PSDF simpler. The orthogonality property of the wavelet basis functions increases computational efficiency, as cross wavelet coefficients are no longer required.

2.5 WAVELET DOMAIN STOCHASTIC RESPONSE OF SDOF SYSTEM

On Fourier transforming Equation (2.18),

$$\hat{\psi}_{a_j,b_i}(\omega)\hat{\psi}^*_{a_k,b_l}(\omega) = \sqrt{a_j a_k}\,\hat{\psi}(a^j\omega)\;\hat{\psi}^*(a^k\omega)e^{i(b_i-b_l)\omega} \tag{2.38}$$

On substituting Equation (2.23) into Equation (2.35),

$$\hat{\psi}_{a_j,b_i}(\omega)\hat{\psi}^*_{a_k,b_l}(\omega) = \frac{\sqrt{a_j a_k}}{2(\sigma-1)\pi}\chi_{\left[\frac{\pi}{a_j},\frac{\sigma\pi}{a_j}\right]}(\omega)\chi_{\left[\frac{\pi}{a_k},\frac{\sigma\pi}{a_k}\right]}(\omega)e^{i(b_i-b_l)\omega} \tag{2.39}$$

The term χ is an indicator function that takes up the value of 1 in the interval $[\cdot]$; else it becomes zero. Using the property of the indicator function, Equation (2.39) may be simplified to

$$\hat{\psi}_{a_j,b_i}(\omega)\hat{\psi}^*_{a_k,b_l}(\omega) = \delta_{jk}\hat{\psi}_{a_j,b_i}(\omega)\hat{\psi}^*_{a_k,b_l}(\omega) \tag{2.40}$$

Let us multiply all terms in Equation (2.33) by $\hat{\psi}^*_{a_k,b_l}(\omega)$ to get the following:

$$\begin{aligned}&\sum_i\sum_j \frac{1}{a_j} W_\psi x_S(a_j,b_i)\hat{\psi}_{a_j,b_i}(\omega)\hat{\psi}^*_{a_k,b_l}(\omega)\\ &= \sum_i\sum_j \frac{1}{a_j} W_\psi \ddot{x}_g(a_j,b_i)H(\omega)\hat{\psi}_{a_k,b_l}(\omega)\hat{\psi}^*_{a_k,b_l}(\omega)\end{aligned} \tag{2.41}$$

Using Equation (2.40), Equation (2.41) may be rewritten as

$$\sum_i W_\psi x_s(a_j,b_i)\hat{\psi}_{a_j,b_i}(\omega) = \sum_i W_\psi \ddot{x}_g(a_j,b_i)H(\omega)\hat{\psi}_{a_j,b_i}(\omega) \tag{2.42}$$

Now, if we take the expectation of the square of the amplitude of both sides of Equation (2.42), integrate over ω and use the orthogonality relationship as defined in Equation (2.37), we will have the following equation:

$$\sum_i E\left[\left|W_\psi x_s(a_j,b_i)\right|^2\right] = \sum_i E\left[\left|W_\psi \ddot{x}_g(a_j,b_i)\right|^2\right]\int_{-\infty}^{\infty}\left|H(\omega)\right|^2\left|\hat{\psi}_{a_j,b_i}(\omega)\right|^2 d\omega$$

$$+\sum_k \sum_{m(\neq k)} E\left[W\ddot{x}_g(a_j,b_k)W\ddot{x}_g^*(a_j,b_m)\right].\int_{-\infty}^{\infty}\hat{\psi}_{a_j,b_k}(\omega)\hat{\psi}^*_{a_j,b_m}(\omega)\left|H(\omega)\right|^2 d\omega$$

(2.43)

The second term on the right-hand side of Equation (2.43) is a complex-conjugate pair, so the same equation may be written as

$$\sum_i E\left[\left|W_\psi x_s(a_j,b_i)\right|^2\right] = \sum_i E\left[\left|W_\psi \ddot{x}_g(a_j,b_i)\right|^2\right]\int_{-\infty}^{\infty}\left|H(\omega)\right|^2\left|\hat{\psi}_{a_j,b_i}(\omega)\right|^2 d\omega$$

$$+\sum_k \sum_{m(\neq k)} E\left[W\ddot{x}_g(a_j,b_k)W\ddot{x}_g^*(a_j,b_m)\right]$$

$$a_j.\int_{-\infty}^{\infty}\left|\hat{\psi}(a^j\omega)\right|^2\left|H(\omega)\right|^2\cos\left\{(b_m-b_k)\omega\right\}d\omega$$

(2.44)

In Equation (2.44), for a specific value of frequency band j and when σ is close to 1, the wavelet coefficients $(W_\psi \ddot{x}_g(a_j,b_k))$ form a narrow-band random process that has slowly modulated amplitudes. Its central frequency is given by $\omega_0 = \frac{\pi}{2a_j}(\sigma+1)$. This corresponds to the central frequency of the jth band of energy. Thus, it may be assumed that the covariance of the wavelet coefficients defined by $E[W\ddot{x}_g(a_j,b_k)W\ddot{x}_g^*(a_j,b_m)]$ as appearing in the second term of the right-hand side of Equation (2.45) contains a term $\cos\{(b_k-b_m)\omega_0\}$ that is modulated by a slowly varying function (say, $S(a_j)$). So, Equation (2.44) may be written as

$$\sum_i E\left[\left|W_\psi x_s(a_j,b_i)\right|^2\right] = \sum_i E\left[\left|W_\psi \ddot{x}_g(a_j,b_i)\right|^2\right]\int_{-\infty}^{\infty}\left|H(\omega)\right|^2\left|\hat{\psi}_{a_j,b_i}(\omega)\right|^2 d\omega$$

$$+\sum_k \sum_{m(\neq k)} a_j S(a_j)\int_{-\infty}^{\infty}\left|\hat{\psi}(a^j\omega)\right|^2\left|H(\omega)\right|^2\cos\{(b_m-b_k)\omega\}.\cos\{(b_k-b_m)\omega_0\}d\omega$$

(2.45)

It may be noted here that because the harmonic oscillations get cancelled, for all $\omega = \omega_0$, the product of two terms, $\cos\{(b_m - b_k)\omega\}$ and $\cos\{(b_k - b_m)\omega_0\}$, in Equation (2.45), when summed over k and m, tends to become zero. Hence, the double summation term in Equation (2.45) may be ignored, and similarly, Equation (2.43) may be simplified to

$$\sum_i E\left[\left|W_\psi x_s(a_j,b_i)\right|^2\right] = \sum_i E\left[\left|W_\psi \ddot{x}_g(a_j,b_i)\right|^2\right]\int_{-\infty}^{\infty}\left|H(\omega)\right|^2\left|\hat{\psi}_{a_j,b_i}(\omega)\right|^2 d\omega \tag{2.46}$$

It is possible to compute the energy (i.e. the mean square response) contribution to the frequency band corresponding to the dilation factor a_j using Parseval's identity and Equations (2.45) and (2.46). The resulting expression for the mean square response then becomes

$$\int_{-\infty}^{\infty} E[|X_s^j(\omega)|^2]d\omega = \sum_i \frac{K\Delta b}{a_j} E\left[\left|W_\psi \ddot{x}_g(a_j,b_i)\right|^2\right]\int_{-\infty}^{\infty}\left|H(\omega)\right|^2\left|\hat{\psi}_{a_j,b_i}(\omega)\right|^2 d\omega \tag{2.47}$$

The unique time-localization property of the wavelet coefficients may now be used to derive from Equation (2.47) the expression for the instantaneous mean square response around $t = b_i$ as follows:

$$\int_{-\infty}^{\infty} E[|X_{si}^j(\omega)|^2]d\omega = \frac{K\Delta b}{a_j} E\left[\left|W_\psi \ddot{x}_g(a_j,b_i)\right|^2\right]\int_{-\infty}^{\infty}\left|H(\omega)\right|^2\left|\hat{\psi}_{a_j,b_i}(\omega)\right|^2 d\omega \tag{2.48}$$

The above expression for the instantaneous mean square response corresponds to the band with dilation factor a_j. As can be seen from Equation (2.48), this is valid only in an integral sense. When σ is quite close to unity, different energy bands become narrower. The integrals in Equation (2.48) have to be performed over a very short time interval, which means that the equality of the integrands and hence a point-wise relation may be safely assumed. For this reason, Equation (2.48) may be rewritten as

$$E[|X_{si}^j(\omega)|^2] = \frac{K\Delta b}{a_j} E\left[\left|W_\psi \ddot{x}_g(a_j,b_i)\right|^2\right]\left|H(\omega)\right|^2\left|\hat{\psi}_{a_j,b_i}(\omega)\right|^2 \tag{2.49}$$

In our case, σ is assumed to be equal to $2^{0.25}$ (i.e. 1.189), and even for this value, we may obtain a reasonable approximation of the instantaneous

PSDF in the case of realistic ground motions. Thus, using the point-wise relation as shown in Equation (2.49) and summing over all energy bands at a particular time instant (i), the instantaneous PSDF of the structural response may be obtained as

$$S_{x_s|_i}(\omega) = \sum_j \frac{E[|X_{si}^j(\omega)|^2]}{\Delta b} = \sum_j \frac{K}{a_j} E\left[\left|W_\psi \ddot{x}_g(a_j, b_i)\right|^2\right] |H(\omega)|^2 \left|\hat{\psi}_{a_j, b_i}(\omega)\right|^2 \tag{2.50}$$

The moments of instantaneous PSDF should be found out to obtain the characteristics of the structural response process, $x_s(t)$. The evaluation of the following statistical parameters would be helpful to obtain more information about the response process.

The nth moment of the instantaneous PSDF, $S_x(\omega)|_{t=b_i}$, is obtained using the following equation:

$$m_n\big|_{t=b_i} = \int_0^\infty \sum_j \frac{K' E\left[\left|W_\psi \ddot{x}_g(a_j, b_i)\right|^2\right] \omega^n \chi_{\left[\frac{\pi}{a_j}, \frac{\sigma\pi}{a_j}\right]}(\omega)}{(\omega^2 - \omega_n^2)^2 + (2\zeta_s \omega\omega_n)^2} d\omega \tag{2.51}$$

In the above expression, $K' = \frac{K}{2(\sigma-1)\pi}$. The closed-form analytical expressions of the zeroth, first, second and fourth moments of instantaneous PSDF are expressed, respectively, as follows:

$$m_0\big|_{t=b_i} = \sum_j K' E\left[\left|W_\psi \ddot{x}_g(a_j, b_i)\right|^2\right] I_{0,j}(\omega_n, \zeta_s) \tag{2.52}$$

$$m_1\big|_{t=b_i} = \sum_j K' E\left[\left|W_\psi \ddot{x}_g(a_j, b_i)\right|^2\right] I_{1,j}(\omega_n, \zeta_s) \tag{2.53}$$

$$m_2\big|_{t=b_i} = \sum_j K' E\left[\left|W_\psi \ddot{x}_g(a_j, b_i)\right|^2\right] I_{2,j}(\omega_n, \zeta_s) \tag{2.54}$$

$$\begin{aligned} m_4\big|_{t=b_i} &= \sum_j K' E\left[\left|W_\psi \ddot{x}_g(a_j, b_i)\right|^2\right] \\ &\times \left[(\sigma - 1)\frac{\pi}{a_j} - \omega_n^4 I_{0,j}(\omega_n, \xi_s) + 2\omega_n^2\left(1 - 2\xi_s^2\right) I_{2,j}(\omega_n, \xi_s)\right] \end{aligned} \tag{2.55}$$

In the above equations, the expressions for the terms $I_{0,j}$, $I_{1,j}$ and $I_{2,j}$ are shown below:

$$I_{0,j} = \int_{\pi/a_j}^{\sigma\pi/a_j} \frac{1}{(\omega^2 - \omega_n^2)^2 + (2\zeta_s \omega \omega_n)^2} d\omega \tag{2.56}$$

$$I_{1,j} = \int_{\pi/a_j}^{\sigma\pi/a_j} \frac{\omega}{(\omega^2 - \omega_n^2)^2 + (2\zeta_s \omega \omega_n)^2} d\omega \tag{2.57}$$

$$I_{2,j} = \int_{\pi/a_j}^{\sigma\pi/a_j} \frac{\omega^2}{(\omega^2 - \omega_n^2)^2 + (2\zeta_s \omega \omega_n)^2} d\omega \tag{2.58}$$

It is worth mentioning here that the instantaneous moments as shown in Equations (2.52) to (2.55) depend on the terms $I_{0,j}$, $I_{1,j}$ and $I_{2,j}$, which in turn are dependent on the natural frequency and the damping ratio of the system under consideration. Thus, the response of the system may get modified considerably by these two basic dynamical properties (frequency and damping) of the system. It is interesting to note that some of the statistical parameters of the process (SDOF system response in this case) that are evaluated using the moments of the PSDF may be used to obtain the ordered peak amplitude of the process. Of these statistical parameters, instantaneous rates of zero crossing, peak occurrence, and bandwidth parameter are important.

2.6 STATISTICAL PARAMETERS AND NONSTATIONARY PEAK RESPONSES

The statistical parameters required to explain the characteristics of a random nonstationary process and structural response are defined below. The instantaneous rate of zero crossings may be defined by

$$\Omega_i = \sqrt{\frac{m_2 \,|_{t=b_i}}{m_0 \,|_{t=b_i}}} \tag{2.59}$$

The instantaneous rate of occurrence of peaks of the response process is defined as follows:

$$N_i = \frac{1}{2\pi}\sqrt{\frac{m_4 \mid_{t=b_i}}{m_2 \mid_{t=b_i}}} \tag{2.60}$$

The instantaneous bandwidth parameter is defined as follows:

$$\epsilon_i = \sqrt{1 - \frac{m_2^2 \mid_{t=b_i}}{m_0 \mid_{t=b_i} m_4 \mid_{t=b_i}}} \tag{2.61}$$

In connection to the first passage problem, which is related to the reliability of the structure when the earthquake excitation and structural response are modelled as random processes [14], another bandwidth parameter, λ_i, is defined as

$$\lambda_i = \sqrt{1 - \frac{m_1^2 \mid_{t=b_i}}{m_0 \mid_{t=b_i} m_2 \mid_{t=b_i}}} \tag{2.62}$$

Now, assume that the structural response process $x_s(t)$ has the duration T and has zeroth instantaneous moment σ_i^2. It may also be assumed that this process, at $t = b_i$, has statistical parameters denoted by Ω, N_i, $\in_i$ and λ_i. The probability that the amplitude of the process, i.e. $|x_s(t)|$ remains below the level x within the time interval $[0, T]$ may be written as [14]

$$P_T(x) = e^{-\int_0^T \alpha(t)dt} = e^{-\sum_i \alpha(t)\mid_{t=b_i} \Delta b} \tag{2.63}$$

where $\alpha(t)$ is the time-dependent rate of the Poisson crossing process and at a particular time instant it is defined as

$$\alpha(t)\mid_{t=b_i} = \frac{\Omega_i}{\pi} e^{-\left(x^2/2\sigma_i^2\right)} \tag{2.64}$$

The probability as defined by Equation (2.63) follows an exponential law. Equation (2.64) gives an approximate estimation of the Poisson crossing process. A better estimation of $\alpha(t)$ is given by the following expression:

$$\alpha(t) = \frac{\Omega_i}{\pi} e^{-\left(x^2/2\sigma_i^2\right)} \frac{1 - exp\left(-\sqrt{\frac{\pi}{2}}\lambda_{ie}(t)\frac{x}{\sigma_i}\right)}{1 - \exp\left(\frac{x^2}{2\sigma_i^2}\right)} \tag{2.65}$$

where $\lambda_{ie} = \lambda_i^{1+b}$ with $b = 0.2$. It might be interesting to note that though the first passage problem is quite suitable for application to nonstationary processes with time-varying statistics, it cannot be used to predict the second-, third-, or higher-order peak amplitudes of the process. However, Gupta and Trifunac [15] proposed order statistics formulation for approximate estimation of higher-order peaks. Basu and Gupta [12] proposed a generalized approach based on this order statistics formulation for an equivalent stationary process for wider applicability of the proposed analysis. This equivalent process is characterized in such a way that the expected amplitude of the largest peak response corresponding to the order statistics formulation is the same as that estimated by using the first passage formulation for $|x(t)|$. The duration of the equivalent process may be calculated from the following equation using the rate of zero crossings, ν_0.

$$T^* = \int_0^T \nu_0(t)\frac{dt}{\nu_0(b_m)} \tag{2.66}$$

The first passage estimate of the largest peak [12] is done from

$$E\left(x_{(1)}\right) = \int_0^T P_T^{-1}(u)du \tag{2.67}$$

The mean peak factor of the ith largest peak amplitude may be subsequently calculated as follows:

$$I_{0,j} = \frac{1}{8\zeta_s\omega_d\omega_n^3}\cdot\left[\zeta_s\omega_n ln\left[\frac{\left\{\left(\frac{\pi\sigma}{a_j}+\omega_d\right)^2+\zeta_s^2\omega_n^2\right\}\left\{\left(\frac{\pi}{a_j}-\omega_d\right)^2+\zeta_s^2\omega_n^2\right\}}{\left\{\left(\frac{\pi\sigma}{a_j}+\omega_d\right)^2+\zeta_s^2\omega_n^2\right\}\left\{\left(\frac{\pi}{a_j}+\omega_d\right)^2+\zeta_s^2\omega_n^2\right\}}\right]\right.$$
$$+2\omega_d\left\{\tan^{-1}\left(\frac{\frac{\pi\sigma}{a_j}+\omega_d}{\zeta_s\omega_n}\right)+\tan^{-1}\left(\frac{\frac{\pi\sigma}{a_j}-\omega_d}{\zeta_s\omega_n}\right)\right. \tag{2.68}$$
$$\left.\left.-\tan^{-1}\left(\frac{\frac{\pi}{a_j}+\omega_d}{\zeta_s\omega_n}\right)-\tan^{-1}\left(\frac{\frac{\pi}{a_j}-\omega_d}{\zeta_s\omega_n}\right)\right\}\right]$$

$$I_{1,j} = \frac{1}{4\zeta_s\omega_d\omega_n} \cdot \left[\tan^{-1}\left(\frac{\frac{\pi}{a_j}+\omega_d}{\zeta_s\omega_n}\right) - \tan^{-1}\left(\frac{\frac{\pi}{a_j}-\omega_d}{\zeta_s\omega_n}\right) - \tan^{-1}\left(\frac{\frac{\sigma\pi}{a_j}+\omega_d}{\zeta_s\omega_n}\right) + \tan^{-1}\left(\frac{\frac{\sigma\pi}{a_j}-\omega_d}{\zeta_s\omega_n}\right)\right] \tag{2.69}$$

$$I_{2,j} = \frac{1}{8\omega_d} ln\left[\frac{\left\{\left(\frac{\sigma\pi}{a_j}-\omega_d\right)^2+\zeta_s^2\omega_n^2\right\}+\left\{\left(\frac{\pi}{a_j}+\omega_d\right)^2+\zeta_s^2\omega_n^2\right\}}{\left\{\left(\frac{\sigma\pi}{a_j}+\omega_d\right)^2+\zeta_s^2\omega_n^2\right\}+\left\{\left(\frac{\pi}{a_j}-\omega_d\right)^2+\zeta_s^2\omega_n^2\right\}}\right] + \frac{1}{4\zeta\omega_n}\left\{\tan^{-1}\left(\frac{\frac{\sigma\pi}{a_j}+\omega_d}{\zeta_s\omega_n}\right) + \tan^{-1}\left(\frac{\frac{\sigma\pi}{a_j}-\omega_d}{\zeta_s\omega_n}\right) - \tan^{-1}\left(\frac{\frac{\pi}{a_j}+\omega_d}{\zeta_s\omega_n}\right) - \tan^{-1}\left(\frac{\frac{\pi}{a_j}-\omega_d}{\zeta_s\omega_n}\right)\right\} \tag{2.70}$$

2.7 WAVELET DOMAIN STOCHASTIC RESPONSE OF MDOF SYSTEM

In this section the formulation of wavelet-based analysis of MDOF systems will be developed. The work by Basu and Gupta [12, 16, 17] in this area is notable, as the main guidelines for the formulation are laid out in these papers. This is necessary, as at certain times a simplified SDOF model does not always represent the model in a true sense. Let us consider an n-storey building structure representing a MDOF system, undergoing seismic excitation, $\ddot{x}_g(t)$, at the base. The floors are all moving with respect to the ground and generically denoted by $x_i(t)$, where i denotes the floor number (varying from 1 to n). The dynamic equilibrium of such a MDOF system may be mathematically expressed by the following equation:

$$[M]\{\ddot{x}_i(t)\}+[C]\{\dot{x}_i(t)\}+[K]\{x_i(t)\}=-[M]\{I_g\}\ddot{x}_g(t) \tag{2.71}$$

The terms $[M]$, $[C]$ and $[K]$ denote mass matrix, damping matrix and stiffness matrix, respectively, each of dimension $[n \times n]$, and I_g is the ground displacement influence vector of size $[n \times 1]$. The floor displacement term, $x_i(t)$, may be expressed as a summation of its normal modes as follows:

$$\{x_i(t)\}=\sum_{p=1}^{n}\{\varnothing^j\}\chi_p(t); \quad p=1,2,3,\ldots \tag{2.72}$$

On taking the wavelet transform of Equation (2.72), the relative displacement at each floor level may be expressed in terms of modal displacements (χ_p) as

$$Wx_i(a_j,b_i)=\sum_{p=1}^{n}\varnothing_i^{(p)}W\chi_p(a_j,b_i) \tag{2.73}$$

The term $\varnothing_i^{(p)}$ indicates the ith element of mode shape $\varnothing_i^{(p)}$. From the last equation, on squaring all terms of Equation (2.73) and then taking expectation, the following equation is obtained, which contains cross-modal terms:

$$E\left[Wx_i(a_j,b_i)\right]^2=\sum_{p=1}^{n}\left[\left(\varnothing_i^{(p)}\right)^2 E\left[W\chi_p(a_j,b_i)\right]^2 + \sum_{l=1,l\neq k}^{n}\varnothing_i^{(k)}\varnothing_i^{(l)}E[W\chi_k(a_j,b_i)W\chi_l(a_j,b_i)]\right] \tag{2.74}$$

On using Equation (2.72), Equation (2.71) may be transformed into n uncoupled equations of motion as

$$\ddot{\chi}_p+2\zeta_p\omega_p\dot{\chi}_p+\omega_p^2\chi_p=-\frac{\{\varnothing_i^{(p)}\}^T[M]\{\Gamma\}}{\{\varnothing_i^{(p)}\}^T[M]\{\varnothing_i^{(p)}\}}\ddot{x}_g; \quad p=1,2,3,\ldots n \tag{2.75}$$

Each of these uncoupled equations represents the vibration of the system in a particular mode. The terms ζ_p and ω_p represent damping ratio and natural frequency, respectively, corresponding to the pth mode. On the right-hand side of Equation (2.75), the fraction $\frac{\{\varnothing_i^{(p)}\}^T[M]\{\Gamma\}}{\{\varnothing_i^{(p)}\}^T[M]\{\varnothing_i^{(p)}\}}$ represents the modal mass participation factor. Now, we need the wavelet expansion of the term x_p, which should be similar to the wavelet expansion of the ground acceleration term $\ddot{x}_g$, as shown in Equation (2.11). So, it may be written in the same way:

$$\chi_p(t) = \sum_i \sum_j \frac{K\Delta b}{a_j} W_\psi \chi_p(a_j, b_i) \psi_{a_j,b_i}(t) \tag{2.76}$$

Substitute Equations (2.11) and (2.76) in Equation (2.75), and then taking Fourier transform, the wavelet coefficients of modal displacements may be expressed in terms of wavelet coefficients of ground motions through the following equation:

$$\sum_i \sum_j \frac{1}{a_j} W\chi_p(a_j, b_i)\hat{\psi}_{a_j,b_i}(\omega) = \sum_i \sum_j \frac{1}{a_j} W\ddot{x}_g(a_j, b_i) T_p(\omega)\hat{\psi}_{a_j,b_i}(\omega) \tag{2.77}$$

The term $T_p(\omega)$ denotes the frequency-dependent complex-valued transfer function at the pth mode correlating the wavelet functionals of the displacement response of the system to the wavelet functionals of the random ground acceleration process and is given as

$$T_p(\omega) = \frac{1}{\left(\omega^2 - \omega_p^2\right) + 2\zeta_p \omega \omega_p} \tag{2.78}$$

Multiplying both sides of Equation (2.77) by $\hat{\psi}^*_{a_j,b_i}(\omega)$, one may get

$$\begin{aligned} &\sum_i \sum_j \frac{1}{a_j} W\chi_p(a_j, b_i)\hat{\psi}_{a_j,b_i}(\omega)\hat{\psi}^*_{a_k,b_l}(\omega) \\ &= \sum_i \sum_j \frac{1}{a_j} W\ddot{x}_g(a_j, b_i) T_p(\omega)\hat{\psi}_{a_j,b_i}(\omega)\hat{\psi}^*_{a_k,b_l}(\omega) \end{aligned} \tag{2.79}$$

Using the relation that $\hat{\psi}_{a_j,b_i}(\omega)(\omega)\hat{\psi}^*_{a_k,b_l}(\omega) = 0$ for $j \neq k$ (which comes from Equation (2.35)), one may further get the following relation for mode p as

$$\sum_j W\chi_p(a_j,b_i)\hat{\psi}_{a_j,b_i}(\omega) = \sum_j W\ddot{x}_g(a_j,b_i)T_p(\omega)\hat{\psi}_{a_j,b_i}(\omega) \tag{2.80}$$

and for mode q as

$$\sum_j W\chi_q(a_j,b_i)\hat{\psi}_{a_j,b_i}(\omega) = \sum_j W\ddot{x}_g(a_j,b_i)T_q(\omega)\hat{\psi}_{a_j,b_i}(\omega) \tag{2.81}$$

Multiply Equation (2.80) by the complex conjugate of Equation (2.81). Subsequently, take expectation of terms on both sides and integrate over ω. On doing so, some cross terms will appear containing coefficients $E[W\ddot{x}_g(a_j,b_k)W\ddot{x}_g(a_j,b_l)]$ $(l \neq k)$, which may be neglected (as they would be zero). Hence, the resulting equation may be obtained as

$$\sum_q E[W\chi_p(a_j,b_i)W\chi_q(a_j,b_i)] = \sum_q \beta_p\beta_q E[W\ddot{x}_g(a_j,b_i)]^2 \int_{-\infty}^{\infty} \left|T_p(\omega)T_q^*(\omega)\right| \left|\hat{\psi}_{a_j,b_i}(\omega)\right|^2 \tag{2.82}$$

When $p = q$, the equation may be simplified to

$$\sum_q E[W\chi_p(a_j,b_i)]^2 = \sum_q \beta_p^2 E[W\ddot{x}_g(a_j,b_i)]^2 \int_{-\infty}^{\infty} \left|T_p(\omega)\right|^2 \left|\hat{\psi}_{a_j,b_i}(\omega)\right|^2 \tag{2.83}$$

Substitute Equations (2.82) and (2.83) in Equation (2.74) to obtain the expected response of the structure in terms of wavelet coefficients of the seismic ground motion process through modal transfer functions as follows:

$$\int_{-\infty}^{\infty} E\left[\left|X_i(\omega)\right|^2\right] d\omega = \int_{-\infty}^{\infty} \sum_i \sum_j \frac{K\Delta b}{a_j} E[W\ddot{x}_g(a_j,b_i)]^2 \sum_{p=1}^{n} \Bigg[\left(\varnothing_i^{(p)}\right)^2 \alpha_p^2 \left|T_p(\omega)\right|^2 + \sum_{q=1,q\neq p}^{n} \varnothing_i^{(p)}\varnothing_i^{(q)}\alpha_j\alpha_k \mathrm{Re}\left(T_p(\omega)T_q^*(\omega)\right) \Bigg] \left|\hat{\psi}_{a_j,b_i}(\omega)\right|^2 d\omega \tag{2.84}$$

The instantaneous mean square value of the ground motion (or any random) process and the expected value of the instantaneous mean square energy of the same process in a frequency band a_j are respectively given by the following equations [16]:

$$E[\ddot{x}_g^2(t)]_{t=b_i} = K\sum_j \frac{E[W\,\ddot{x}_g(a_j,b_i)]^2}{a_j} \tag{2.85}$$

$$\frac{1}{\Delta b}\int_{-\infty}^{\infty} E\left[\left|X_i^j(\omega)\right|^2\right] d\omega = \frac{K}{a_j} E[W\,\ddot{x}_g(a_j,b_i)]^2 \tag{2.86}$$

The instantaneous mean square value of the structural response $x_i(t)$ and its instantaneous mean square energy at time instant b_q in the frequency band a_p may be similarly written as

$$E\left[x_i^2(t)\right]\Big|_{t=b_i} = K\sum_p \frac{E\left[W\ddot{x}_g(a_p,b_q)\right]^2}{a_p} \int_{-\infty}^{\infty} E\left[W\ddot{x}_g(a_p,b_q)\right]^2 \\ \times \sum_{j=1}^{n}\left[\left(\phi_i^{(j)}\right)^2 \alpha_j^2 \left|H_j(\omega)\right|^2 + \sum_{k=1.k\neq j}^{n} \phi_i^{(j)}\phi_i^{(k)}\alpha_j\alpha_k \mathrm{Re}(H_j(\omega)H_k^*(\omega))\right] \\ \times \left|\hat{\psi}_{a_p,b_q}(\omega)\right|^2 d\omega \tag{2.87}$$

$$\int_{-\infty}^{\infty} E\left[\left|X_i^p(\omega)\right|^2\right] d\omega = \int_{-\infty}^{\infty} \frac{K\Delta b}{a_p} E[W\,\ddot{x}_g(a_p,b_q)]^2 \sum_{j=1}^{n}\left[(\phi_i^{(j)})^2\alpha_j^2\left|H_j(\omega)\right|^2 \right. \\ \left. + \sum_{k=1.k\neq j}^{n} \phi_i^{(j)}\phi_i^{(k)}\alpha_j\alpha_k \mathrm{Re}(H_j(\omega)H_k^*(\omega))\right] \left|\hat{\psi}_{a_p,b_q}(\omega)\right|^2 d\omega \tag{2.88}$$

Equations (2.87) and (2.88) are true in an integral sense. However, if σ is chosen close to the value of 1, the relationships may be true in a point-wise sense too. It has been seen from much modelling of nonstationary processes based on wavelets that a choice of $\sigma = 2^{1/4}$ is quite a good approximation.

Hence, on using this value of σ, Equation (2.88) may be discretized to the following form:

$$E\left[\left|X_i^p(\omega)\right|^2\right] = \frac{K\Delta b}{a_p} E[W\ddot{x}_g(a_p,b_q)]^2 \sum_{j=1}^{n}\left[\left(\phi_i^{(j)}\right)^2 \alpha_j^2 \left|H_j(\omega)\right|^2 + \sum_{k=1.k\neq j}^{n} \phi_i^{(j)}\phi_i^{(k)}\alpha_j\alpha_k \mathrm{Re}\left(H_j(\omega)H_k^*(\omega)\right)\right]\left|\hat{\psi}_{a_p,b_q}(\omega)\right|^2 d\omega \tag{2.89}$$

As we have done earlier in the case of a SDOF system, summing up Equation (2.89) over all energy bands (p) and then averaging over time interval Δb, the instantaneous power spectral density function at a specific time instant ($t = b_q$) may be written as

$$S_{x_i}^q = \sum_p \frac{E\left[\left|X_i^p\right|^2\right]}{\Delta b} = \sum_p \frac{K}{a_p} E\left[\left|Wz\left(a_p,b_q\right)\right|^2\right] \times \sum_{j=1}^{n}\left[\sum_{j=1}^{n}\left[\left(\phi_i^{(j)}\right)^2 \alpha_j^2\left|H_j(\omega)\right|^2 + \sum_{k=1.k\neq j}^{n} \phi_i^{(j)}\phi_i^{(k)}\alpha_j\alpha_k \mathrm{Re}(H_j(\omega)H_k^*(\omega))\right]\left|\hat{\psi}_{a_p,b_q}(\omega)\right|^2\right] \tag{2.90}$$

The instantaneous PSDF has the following general expression:

$$S_r^q = \sum_p \frac{K}{a_p} E\left[\left|Wz\left(a_p,b_q\right)\right|^2\right] \times \sum_{j=1}^{n}\left[\sum_{j=1}^{n}\left[\left\{\rho_j^2\alpha_j^2 + \sum_{k=1.k\neq j}^{n} \rho_j\rho_k\alpha_j\alpha_k(C_{jk}+D_{jk})\right\}\left|H_j(\omega)\right|^2 - \frac{\left|\omega H_j(\omega)\right|^2}{\omega_j^2}\sum_{k=1.k\neq j}^{n} \rho_j\rho_k\alpha_j\alpha_k D_{jk}\right]\left|\hat{\psi}_{a_p,b_q}(\omega)\right|^2\right] \tag{2.91}$$

The term ρ_j may vary depending on the type of the response. For example, for displacement response this may be denoted by $\phi_i^{(j)}$. The expressions for the two coefficients C_{jk} and D_{jk} are, respectively [16],

$$C_{jk} = \frac{1}{B_{jk}}\left[8\zeta_j\left(\zeta_j + \zeta_k\frac{\omega_k}{\omega_j}\right)\left\{\left(1-\left(\frac{\omega_k}{\omega_j}\right)^2\right)^2 - 4\frac{\omega_k}{\omega_j}\left(\zeta_j - \zeta_k\frac{\omega_k}{\omega_j}\right)\left(\zeta_k - \zeta_j\frac{\omega_k}{\omega_j}\right)\right\}\right] \tag{2.92}$$

$$D_{jk} = \frac{1}{B_{jk}}\left[2\left(1-\left(\frac{\omega_k}{\omega_j}\right)^2\right)\left\{4\frac{\omega_k}{\omega_j}\left(\zeta_j - \zeta_k\frac{\omega_k}{\omega_j}\right)\left(\zeta_k - \zeta_j\frac{\omega_k}{\omega_j}\right) - \left(1-\left(\frac{\omega_k}{\omega_j}\right)^2\right)^2\right\}\right] \tag{2.93}$$

$$B_{jk} = 8\left(\frac{\omega_k}{\omega_j}\right)^2\left[\left(\zeta_j^2 + \zeta_k^2\right)\left(1-\left(\frac{\omega_k}{\omega_j}\right)^2\right)^2 - 2\left(\zeta_k^2 - \zeta_j^2\left(\frac{\omega_k}{\omega_j}\right)^2\right)\left(\zeta_j^2 - \zeta_k^2\left(\frac{\omega_k}{\omega_j}\right)^2\right)\right] + \left(1-\left(\frac{\omega_k}{\omega_j}\right)^2\right)^4 \tag{2.94}$$

Chapter 3

Ground motion characterization and PSA response of SDOF system

In Chapter 2, the theoretical background for obtaining a nonstationary stochastic response of a single-degree-of-freedom (SDOF) system subjected to a random vibration process in the wavelet domain is clearly shown. The formulation to obtain peak stochastic responses of a system is elaborated in the same chapter. Users should keep in mind that to study the response of a system subjected to nonstationary ground motions, it is necessary to characterize the seismic process through statistical functionals of wavelet coefficients. The ground motion characterization has been attempted before by researchers using different approaches like spectrum-compatible wavelet functionals for the input Fourier spectrum and input response spectrum. In any ground motion, peak horizontal ground acceleration is important, as it is directly related to the damage potential. However, the duration of strong shaking, the energy and frequency contents, peak ground velocities and displacements, etc. also affect the extent of damage in structures. The frequency content of earthquake ground motion is generally characterized by the shape of the acceleration response spectrum. The energy content in the spectrum is directly proportional to the square of the acceleration. In this chapter, we will see how the wavelet coefficients generated from a single ensemble are used to characterize ground motions, and how this characterization is subsequently used to obtain the pseudospectral acceleration (PSA) response spectrum of a SDOF system.

3.1 CHARACTERIZATION OF GROUND MOTIONS

The SYNACC program [18–22] is used to calculate wavelet coefficients as per the formulation laid out in Chapter 2. The wavelet-based coefficients are calculated only from a single realization of the synthetic ground motion (accelerogram) process. This nonstationary process represents the earthquake motion recorded near Coyote Hills (Dumbarton Bridge site) during the Loma Prieta earthquake (1989). The same parameters as used by Gupta and Trifunac [15] are considered for generating the synthetic accelerogram for Loma Prieta seismic motion. The wavelet coefficients are calculated for 2047 time points and for each frequency band from –17 to +4. The time points are considered for each of the 22 frequency bands. It means that for a particular frequency band, wavelet coefficients over the entire duration of the seismic excitation (which is covered by 2047 time points) are calculated. Thus, a total of 45 034 wavelet coefficients are calculated. It must be mentioned here that the value of the discretization parameter, Δb, is assumed to be 0.02. The seismic ground acceleration due to the Loma Prieta earthquake is shown in Figure 1.2.

For each frequency band the total duration of seismic excitation T may be divided into a number of smaller stretches for the purpose of local averaging. Each such stretch has a duration of T_j that corresponds to the central frequency of the band considered. It is now known that the range of frequency for the jth frequency band is from $\frac{\pi}{a_j}$ to $\frac{\pi\sigma}{a_j}$. The central frequency, ω_j, for this band is therefore $\frac{1}{2}(\frac{\pi}{a_j}+\frac{\pi\sigma}{a_j})$ rad/s, and hence the duration, T_j, of the stretch corresponding to this central frequency is given as follows:

$$T_j = \frac{2\pi}{\omega_j} = \frac{4a_j}{1+\sigma} \tag{3.1}$$

In each stretch the wavelet coefficients are first squared and subsequently averaged over T_j to represent the value of the energy $E[W_\psi^2 \ddot{x}_g(a_j,b_i)]$ distributed uniformly over the small stretch with duration T_j. The expression for this energy over the stretch may be given as

$$E\left[W_\psi^2\, \ddot{x}_g(a_j,b_i)\right] = \frac{\Delta b}{T_j}\sum_{k_1}^{k_2} W_\psi^2\, \ddot{x}_g(a_j,b_i) \tag{3.2}$$

where

$$k_1 = \frac{4a_j m \Delta b}{1+\sigma} \tag{3.3}$$

and

$$k_2 = \frac{4a_j (m+1)\Delta b}{1+\sigma} \tag{3.4}$$

It may be mentioned here that $k_1 \le i \le k_2$. The squares of the wavelet coefficients computed using the above-mentioned ground motion characterization technique corresponding to the random ground motion process

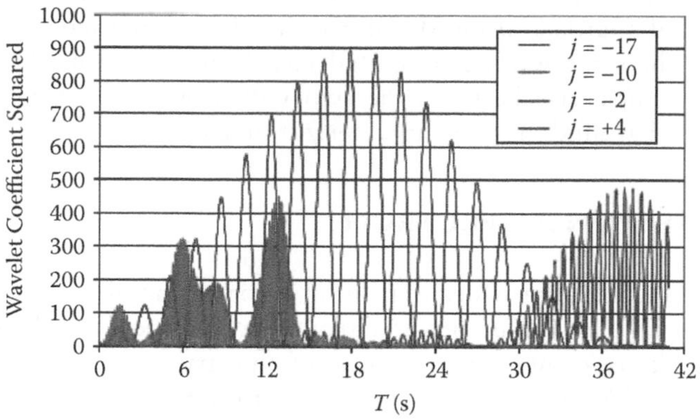

Figure 3.1 Wavelet coefficients for four different frequency bands using Δb = 0.02 s.

shown in Figure 1.2 are plotted in Figure 3.1 for different frequency bands (j = −17, −10, −2 and 4). These coefficients are computed using Δb = 0.02 s. The same coefficients are computed assuming nonorthogonal properties (i.e. $\Delta b = \frac{\sigma^j}{j-1}$) for different frequency bands using the same ground motion and are demonstrated in Figure 3.2. However, in the text and for all examples used in this book, we will use Δb = 0.02 s wherever applicable.

Thus, the wavelet coefficients are used to characterize the ground motions. However, validation should be done to see that this wavelet-based ground motion characterization works satisfactorily. For this purpose,

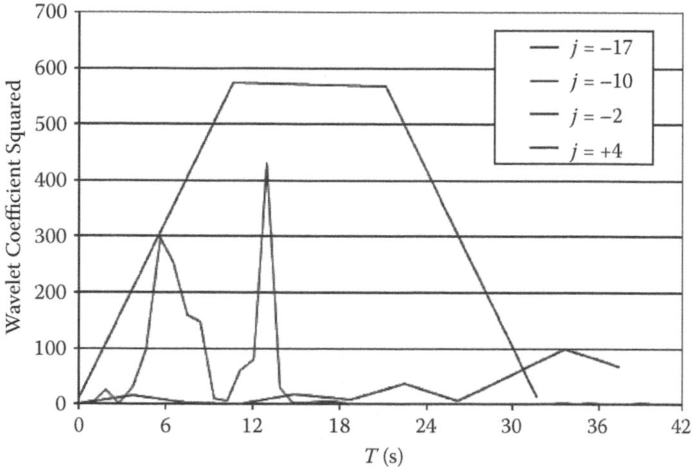

Figure 3.2 Wavelet coefficients for four different frequency bands using $\Delta b = \frac{\sigma^j}{j-1}$.

a pseudospectral acceleration spectrum will now be obtained using the wavelet-based approach using the above-mentioned characterization technique, and the same will be computed based on direct time history simulation. A comparison between the two will reveal the efficiency of the wavelet-based ground motion characterization technique.

3.2 PSA RESPONSE SPECTRUM

A response spectrum provides important information about the influence of an earthquake on a structure. If a ground acceleration time history from an earthquake is known, the response of the structure may be obtained using Newmark's method. This gives rise to a response spectrum. The system equation representing dynamic equilibrium for a SDOF system, as shown in Figure 3.3, subjected to an earthquake excitation is given below:

$$m\ddot{x}(t) + c\dot{x}(t) + kx(t) = -m\ddot{x}_g(t) \tag{3.5}$$

In the above equation, m, c and k denote the mass, the damping coefficient and the stiffness coefficient of the system, and the term $\ddot{x}_g(t)$ denotes the ground acceleration. The term $x(t)$ denotes the displacement of the mass with respect to the free field (ground), and the dots represent derivatives with respect to time. The damping and stiffness coefficients depend on the mass, the damping ratio (ζ) and the natural frequency (ω_n) of the system, as shown in Equations (2.13) and (2.14), respectively. Thus, on substituting these two equations and then simplifying, Equation (3.5) may be rewritten in the following form:

$$\ddot{x}(t) + 2\zeta\omega_n\dot{x}(t) + \omega_n^2 x(t) = -\ddot{x}_g(t) \tag{3.6}$$

The accelerogram contains the values of accelerations at different time points covering the whole duration of the earthquake. Thus, assuming a specific value of the damping ratio (ζ) and for a specific

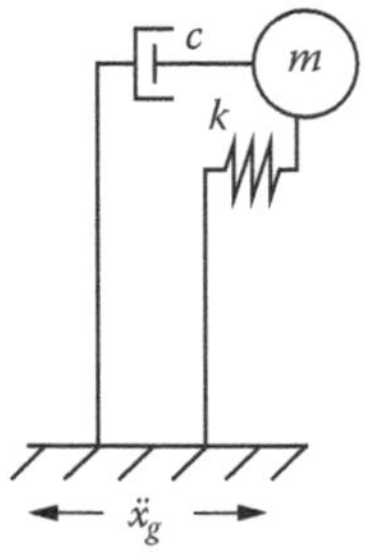

Figure 3.3 Model of a SDOF system.

system frequency (ω_n), the above equation may be solved at each time point with the corresponding value of ground acceleration. This will ultimately generate a row of values of accelerations for all discrete time points considered from which the largest value (the peak) may be easily identified. The process may be reiterated with other different values of system frequency maintaining a constant damping ratio. In the end, a two-dimensional array will be obtained where at a particular damping ratio, for each system frequency there will be a specific value of the system response, $x(t)$. This SDOF system displacement response may now be multiplied by the square of the system frequency (i.e. ω_n^2) to get the pseudoacceleration. The pseudoacceleration values thus obtained for different system frequencies may be plotted against corresponding frequencies to obtain the PSA spectrum. Thus, the PSA spectrum is obtained for a SDOF system subjected to a specific seismic motion at a particular site. The solution procedure follows a time domain approach. The site may undergo several earthquakes, and each ground shaking may be digitally recorded. For each of these recorded accelerograms, a PSA spectrum for the same SDOF system may be computed. The time history simulation results for different accelerograms may then be averaged to obtain a single PSA spectrum for a specific damping ratio.

So far, the thing looks fine. However, the most difficult part is to get so many accelerograms at a particular site, as these may not be readily available. In addition, performing time history simulations to obtain a PSA spectrum for each of these time history acceleration data may be time-consuming. The concept of wavelet transform may come in very handy in this respect. Let us now use the concept of wavelet analysis to obtain a PSA spectrum for a single accelerogram. The displacement response, $x(t)$, of the structure and its velocity and acceleration terms, $\dot{x}(t)$ and $\ddot{x}(t)$, may be expressed respectively in terms of wavelet coefficients following Equation (2.17) as

$$x(t) = \sum_i \sum_j \frac{K\Delta b}{a_j} W_\psi x(a_j, b_i) \psi_{a_j,b_i}(t) \tag{3.7}$$

$$\dot{x}(t) = \sum_i \sum_j \frac{K\Delta b}{a_j} W_\psi x_s(a_j, b_i) \dot{\psi}_{a_j,b_i}(t) \tag{3.8}$$

$$\ddot{x}(t) = \sum_i \sum_j \frac{K\Delta b}{a_j} W_\psi x_s(a_j, b_i) \ddot{\psi}_{a_j,b_i}(t) \tag{3.9}$$

Following a similar procedure as shown in Section 2.3, the wavelet domain equation of motion for the SDOF system is obtained as

$$\sum_i \sum_j \frac{1}{a_j} W_\psi x(a_j, b_i) \hat{\psi}_{a_j, b_i}(\omega) = \sum_i \sum_j \frac{1}{a_j} W_\psi \ddot{x}_g(a_j, b_i) \times \left\{ -\frac{1}{(\omega_n^2 - \omega^2) + i(2i\zeta\omega\omega_n)} \right\} \hat{\psi}_{a_j, b_i}(\omega) \tag{3.10}$$

Multiplying both sides of Equation (3.10) by $\hat{\psi}^*_{a_k, b_l}(\omega)$ and using Equation (2.40), the following equation is obtained:

$$\sum_i W_\psi x(a_j, b_i) \hat{\psi}_{a_j, b_i}(\omega) = \sum_i W_\psi \ddot{x}_g(a_j, b_i) \left\{ -\frac{1}{(\omega_n^2 - \omega^2) + i(2i\zeta\omega\omega_n)} \right\} \hat{\psi}_{a_j, b_i}(\omega) \tag{3.11}$$

Equation (3.11) related the structural displacement response to the ground acceleration in terms of wavelet coefficients. On following Equations (2.43) to (2.49), the instantaneous mean square response (at $t = b_i$) is expressed as

$$E[|x_i^j(\omega)|^2] = \frac{K\Delta b}{a_j} E\left[\left| W_\psi \ddot{x}_g(a_j, b_i) \right|^2 \right] \left| \frac{1}{(\omega_n^2 - \omega^2) + i(2i\zeta\omega\omega_n)} \right|^2 \left| \hat{\psi}_{a_j, b_i}(\omega) \right|^2 \tag{3.12}$$

Using a point-wise relation as shown in Equation (3.12) and obtaining the sum over all energy bands at a particular time instant, the instantaneous PSDF of the SDOF response may be obtained as

$$S_{x|_i}(\omega) = \sum_j \frac{E[|X_i^j(\omega)|^2]}{\Delta b} = \sum_j \frac{K}{a_j} E\left[\left| W_{\psi \ddot{x} g}(a_j, b_i) \right|^2 \right] |H(\omega)|^2 |\, \hat{\psi}_{a_j, b_i}(\omega)|^2 \tag{3.13}$$

Using Equation (2.51), the zeroth, first and second moments of instantaneous PSDF are obtained, from which the largest peak factor of the response is obtained following the guidelines as shown in Section 2.6. The expected response spectra for pseudospectral acceleration may be obtained by using the formulation narrated so far on the basis of expected energy, $[|W_\psi \ddot{x}_g(a_j, b_i)|^2]$. Similar results have also been obtained from exact time history simulation of the same SDOF system using the Runge–Kutta fourth-order method. The comparison of the results between time history simulation and the wavelet-based response spectrum is of particular interest. It must be mentioned here that the PSA spectrum is generated based on statistical functionals of wavelet coefficients calculated from a single accelerogram at a particular site. On the other hand, the PSA curve is generated from time history simulations of 17 different accelerograms at the same site. The PSA spectra thus generated from two different techniques will be compared. However, it would be appropriate to clarify how the exact time history simulation has been performed.

3.3 TIME HISTORY SIMULATION BY THE RUNGE–KUTTA FOURTH-ORDER METHOD

Runge–Kutta (R-K) methods for approximately solving ordinary differential equations primarily represent implicit and explicit iterative methods. R-K methods use information from a single preceding point to estimate the value of the dependent variable of the system (like the system response) for initial value problems.

The general form of any extrapolation equation is given as follows, in which m represents the weighted averages of the slopes at various points in the interval h.

$$y_{i+1} = y_i + mh \tag{3.14}$$

Now, suppose the term m is estimated using slopes at r number of points in the interval (x_i, x_{i+1}). In that case, m may be written as

$$m = n_1 m_1 + n_2 m_2 + n_3 m_3 + \cdots + n_r m_r \tag{3.15}$$

where n_1, n_2, ... n_r are the weights of the gradients or slopes at various points. The gradients m_1, m_2, m_3, ... are calculated as per the following generic term:

$$m_r = f(x_i + a_{r-1}h, y_i + b_{r-1,1}m_1 h + \cdots + b_{(r-1),(r-1)}m_{r-1}h) \tag{3.16}$$

From Equation (3.16) it is clearly seen that for $r = 1$, $m_1 = f(x_i, y_i)$. However, for other values of r the expression becomes

$$m_r = f\left(x_i + a_{r-1}h, y_i + h\sum_{j=1}^{r-1} b_{r-1,j}m_j\right); \quad r \geq 2 \tag{3.17}$$

The slopes are computed recursively using Equation (3.17) starting from m_1. It may be noted here that the slope at any point is computed using slopes at all previous points. The R-K methods are usually known by their order. The r-order R-K method means the slopes at r points are used to construct the weighted average slope, m. Euler's method is a first-order R-K method because only one slope at (x_i, y_i) is used to estimate y_{i+1}. Usually, the higher the order, the better is the accuracy of the estimate. The classical fourth-order R-K method is given below.

The first four gradients have the following expressions:

$$m_1 = f(x_i, y_i) \tag{3.18}$$

$$m_2 = f\left(x_i + \frac{h}{2}, y_i + \frac{m_1 h}{2}\right) \tag{3.19}$$

$$m_3 = f\left(x_i + \frac{h}{2}, y_i + \frac{m_2 h}{2}\right) \tag{3.20}$$

$$m_4 = f(x_i + h, y_i + m_3 h) \tag{3.21}$$

Using these four gradients, y_{i+1} may be evaluated as follows:

$$y_{i+1} = y_i + \left(\frac{m_1 + 2m_2 + 2m_3 + m_4}{6}\right) h \tag{3.22}$$

The fourth-order R-K method is quite accurate. It resembles Simpson's one-third rule of integration. The local error in this method is of the order h^5 and the global error is of the order h^4.

The time history simulation of the SDOF system is carried out 17 times to obtain 17 different PSA spectra corresponding to 17 different accelerograms. In each case the largest peak value of the displacement response is obtained and multiplied by the square of the natural frequency of the SDOF system to obtain the pseudoacceleration term.

3.4 FIXED-BASE ANALYSIS OF A TANK

As a simplistic approach to a more realistic problem, we now confine our attention to the analysis of liquid storage tanks. The response of fluid storage tanks subjected to earthquake ground excitations has been the subject of interest to many researchers and engineers in the last few decades. Such tanks are quite common in nuclear industry, oil and petrochemical industries, sewerage plants, etc. Several academicians and engineers [23–27] in the past have studied the response of liquid storage tanks both experimentally and theoretically without soil–structure interaction effects being considered. We will first formulate the theory of responses of a fixed-base liquid storage tank subjected to ground shaking. This implies that no soil–structure interaction effects are considered, as the tank base is rigidly anchored to the ground. A basic guideline about wavelet domain analysis of a tank is given by Chatterjee and Basu [28]. The wavelet domain formulation of the system equation and its solution are described in steps in the following sections.

3.4.1 Basic assumptions

Let us assume a right circular, cylindrical tank structure containing water filled to its height. The tank radius and its height (which is also the height of the water within the tank) are respectively denoted by R and H. The tank wall is thin and denoted by h. Let the tank be assumed to be placed on a thick foundation and the foundation be supported rigidly on the soil surface, which makes soil–structure interaction effects redundant and hence not considered. In fact, this assumption simplifies the whole analysis and helps us obtain the wavelet-based stochastic response spectrum for the purpose of validation quite easily. The tank–foundation is considered thick to avoid any rocking effects in motion. The tank is rigidly clamped to its base of the same radius as that of the tank. The Young's modulus, Poisson's ratio and mass density of the tank are denoted by E, ν and ρ, respectively. The tank is subjected to a seismic ground motion. The liquid in the tank is assumed to undergo vibrations in impulsive and convective modes. The convective mode of liquid vibration, however, is safely assumed to contain only the first sloshing mode. Hence, the damping ratio and the natural frequency of impulsive liquid vibration (when a part of the liquid mass is vibrating in unison with the tank) are denoted by ξ_i and ω_i. The same terms, in case of convective mode of liquid vibration, are denoted by ξ_c and ω_c. The tank is subjected to a horizontal free-field seismic absolute ground displacement, $x_g(t)$. The impulsive and convective masses of the vibrating liquid are assumed to be lumped at suitable heights above the base (h_i and h_c, respectively). The displacements of the impulsive and the convective masses with respect to the foundation are denoted by $x_i(t)$ and $x_c(t)$, respectively. The front and top views of the tank–foundation are shown in Figure 3.4. The mass-spring-dashpot model representing the tank–liquid–foundation system attached to a rigid base is shown in Figure 3.5.

3.4.2 Equations of motion

From the basic assumptions as stated above and in Figure 3.4, it is clear that the model represents a two-degree-of-freedom (2-DOF) system. In Figure 3.4, the terms C_{di} and C_{dc} denote the damping ratios of impulsive

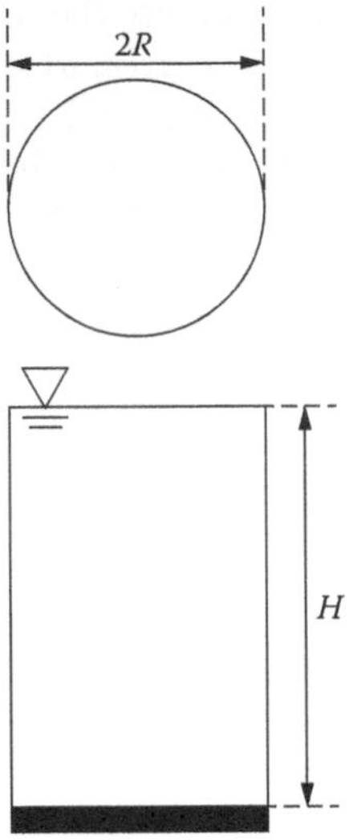

Figure 3.4 Top and front views of the tank–liquid system.

and convective modes of liquid mass, the terms K_i and K_c denote the stiffness coefficients of impulsive and convective liquid vibration modes, and the terms m_i and m_c represent the masses of impulsive and convective modes, respectively. The equation of motion considering the impulsive mass of the tank–liquid may be written as

$$m_i\,\ddot{x}_i(t) + m_i\omega_i^2 x_i(t) + 2m_i\zeta_i\omega_i\dot{x}_i(t) = -m_i\,\ddot{x}_g(t) \tag{3.23}$$

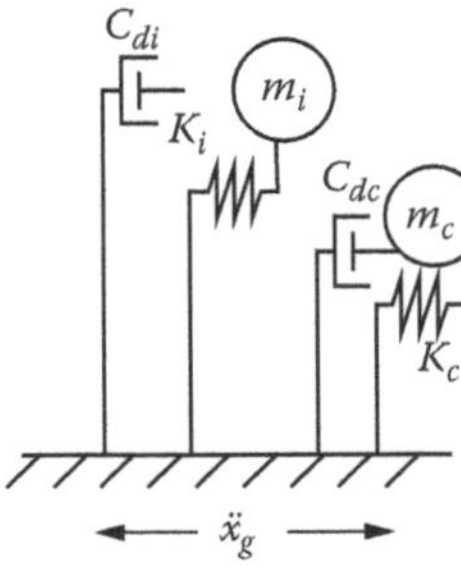

Figure 3.5 2-DOF fixed-base tank–liquid system.

Table 3.1 Values of C_i for different ratios of tank wall thickness to tank radius and tank height to tank radius

h/R	H/R = 0.5	H/R = 1.0	H/R = 2.0	H/R = 3.0
0.0005	0.0506	0.0620	0.0637	0.0563
0.001	0.0719	0.0875	0.0896	0.0792
0.002	0.1019	0.1231	0.1254	0.1108

Source: Veletsos, A. S., and Tang, Y., *Earthquake Eng. Struct. Dyn.*, 19, 473–496, 1990.

Similarly, the equation of motion considering the vibration of the convective mass of the tank–liquid system may be written as

$$m_c\,\ddot{x}_c(t) + m_c\omega_c^2 x_c(t) + 2m_c\zeta_c\omega_c\dot{x}_c(t) = -m_c\,\ddot{x}_g(t) \tag{3.24}$$

The first (impulsive) mode of liquid vibration has the frequency ω_i expressed as follows [27]:

$$\omega_i = \frac{C_i}{H}\sqrt{\frac{E}{\rho}} \tag{3.25}$$

The term C_i is a dimensionless coefficient that depends on certain parameters, like the ratio of the depth of the liquid in the tank to the radius of the tank (i.e. H/R), the ratio of wall thickness to the tank radius (i.e. h/R) and the Poisson's ratio for the tank material, the ratio of the mass density of the liquid to that of the tank (i.e. $\frac{\rho_l}{\rho}$). Table 3.1 shows the values of C_i for different values of h/R and H/R.

The frequency, ω_c, of the first sloshing mode of vibration of the convective mass of the tank–liquid is expressed as follows [27]:

$$\omega_c = 2\pi\sqrt{\lambda_1 \tanh\left(\lambda_1 \frac{H}{R}\right)\frac{g}{R}} \tag{3.26}$$

In the expression of ω_c, g represents acceleration due to gravity and λ_1 is the first root of the first derivative of the Bessel function of first kind and first order. On simplifying Equations (3.23) and (3.24) we get the following equations:

$$\ddot{x}_i(t) + \omega_i^2 x_i(t) + 2\zeta_i\omega_i\dot{x}_i(t) = -\ddot{x}_g(t) \tag{3.27}$$

and

$$\ddot{x}_c(t) + \omega_c^2 x_c(t) + 2\zeta_c\omega_c\dot{x}_c(t) = -\ddot{x}_g(t) \tag{3.28}$$

3.4.3 Wavelet domain formulation

The wavelet representation of the ground motion is given in Equation (2.11). As we have seen earlier in Chapter 2, on differentiating the expressions of impulsive and convective displacements of the liquid mass, the velocities and accelerations may be obtained as follows:

$$x_i(t) = \sum_i \sum_j \frac{K' \Delta b}{a_j} W_\psi x_i(a_j, b_i) \psi_{a_j, b_i}(t) \tag{3.29}$$

$$\dot{x}_i(t) = \sum_i \sum_j \frac{K \Delta b}{a_j} W_\psi x_i(a_j, b_i) \dot{\psi}_{a_j, b_i}(t) \tag{3.30}$$

$$\ddot{x}_i(t) = \sum_i \sum_j \frac{K \Delta b}{a_j} W_\psi x_i(a_j, b_i) \ddot{\psi}_{a_j, b_i}(t) \tag{3.31}$$

$$x_c(t) = \sum_i \sum_j \frac{K \Delta b}{a_j} W_\psi x_c(a_j, b_i) \psi_{a_j, b_i}(t) \tag{3.32}$$

$$\dot{x}_c(t) = \sum_i \sum_j \frac{K \Delta b}{a_j} W_\psi x_c(a_j, b_i) \dot{\psi}_{a_j, b_i}(t) \tag{3.33}$$

$$\ddot{x}_c(t) = \sum_i \sum_j \frac{K \Delta b}{a_j} W_\psi x_c(a_j, b_i) \ddot{\psi}_{a_j, b_i}(t) \tag{3.34}$$

On substituting wavelet-based expressions of $x_i(t)$ and $x_c(t)$ and their derivatives from Equations (3.29) to (3.34) in Equation (3.27) and (3.28), respectively, one may obtain the following equations:

$$\begin{aligned} &\sum_i \sum_j \frac{K \Delta b}{a_j} W_\psi x_i(a_j, b_i) \ddot{\psi}_{a_j, b_i}(t) + 2\zeta_i \omega_i \sum_i \sum_j \frac{K \Delta b}{a_j} W_\psi x_i(a_j, b_i) \dot{\psi}_{a_j, b_i}(t) \\ &+ \omega_i^2 \sum_i \sum_j \frac{K \Delta b}{a_j} W_\psi x_i(a_j, b_i) \psi_{a_j, b_i}(t) = -\sum_i \sum_j \frac{K \Delta b}{a_j} W_\psi \ddot{x}_g(a_j, b_i) \psi_{a_j, b_i}(t) \end{aligned} \tag{3.35}$$

$$\sum_i\sum_j \frac{K\Delta b}{a_j} W_\psi x_c(a_j,b_i)\ddot{\psi}_{a_j,b_i}(t) + 2\zeta_c\omega_c \sum_i\sum_j \frac{K\Delta b}{a_j} W_\psi x_c(a_j,b_i)\dot{\psi}_{a_j,b_i}(t)$$
$$+\omega_c^2 \sum_i\sum_j \frac{K\Delta b}{a_j} W_\psi x_c(a_j,b_i)\psi_{a_j,b_i}(t) = -\sum_i\sum_j \frac{K\Delta b}{a_j} W_\psi\, \ddot{x}_g(a_j,b_i)\psi_{a_j,b_i}(t) \tag{3.36}$$

On simplifying further, as shown in Section 2.3, the above equations may be further simplified to obtain the following two equations in the time domain:

$$\sum_i\sum_j \frac{1}{a_j} W_\psi x_i(a_j,b_i)\left\{\ddot{\psi}_{a_j,b_i}(t) + 2\zeta_i\omega_i\dot{\psi}_{a_j,b_i}(t) + \omega_i^2\psi_{a_j,b_i}(t)\right\}$$
$$= -\sum_i\sum_j \frac{1}{a_j} W_\psi\, \ddot{x}_g(a_j,b_i)\psi_{a_j,b_i}(t) \tag{3.37}$$

$$\sum_i\sum_j \frac{1}{a_j} W_\psi x_c(a_j,b_i)\left\{\ddot{\psi}_{a_j,b_i}(t) + 2\zeta_c\omega_c\dot{\psi}_{a_j,b_i}(t) + \omega_c^2\psi_{a_j,b_i}(t)\right\}$$
$$= -\sum_i\sum_j \frac{1}{a_j} W_\psi\, \ddot{x}_g(a_j,b_i)\psi_{a_j,b_i}(t) \tag{3.38}$$

On performing Fourier transform on time domain equations as obtained above and collecting proper terms on both sides, the displacements corresponding to impulsive and convective modes of vibration may be expressed in terms of ground acceleration using wavelet coefficients as

$$\sum_i\sum_j \frac{1}{a_j} W_\psi x_i(a_j,b_i)\hat{\psi}_{a_j,b_i}(\omega) = \sum_i\sum_j \frac{1}{a_j} W_\psi \ddot{x}_g(a_j,b_i)H_i(\omega)\hat{\psi}_{a_j,b_i}(\omega) \tag{3.39}$$

$$\sum_i\sum_j \frac{1}{a_j} W_\psi x_c(a_j,b_i)\hat{\psi}_{a_j,b_i}(\omega) = \sum_i\sum_j \frac{1}{a_j} W_\psi \ddot{x}_g(a_j,b_i)H_i(\omega)\hat{\psi}_{a_j,b_i}(\omega) \tag{3.40}$$

The terms $H_i(\omega)$ and $H_c(\omega)$ are the frequency-dependent transfer functions relating the wavelet coefficients of the relative horizontal displacements x_i of the impulsive liquid mass and x_c of the convective liquid mass to the wavelet coefficients of the ground acceleration. These transfer functions have the following expressions:

$$H_i(\omega) = -\frac{1}{(\omega_i^2 - \omega^2) + i(2\zeta_i\omega\omega_i)} \tag{3.41}$$

$$H_c(\omega) = -\frac{1}{(\omega_c^2 - \omega^2) + i(2\zeta_c\omega\omega_c)} \tag{3.42}$$

On multiplying Equations (3.39) and (3.40) by $\hat{\psi}^*_{a_k,b_l}(\omega)$ and using Equation (2.40), the sum of the wavelet coefficients of displacements of impulsive and convective liquid masses over time may be expressed in terms of wavelet functionals of ground motions as follows:

$$\sum_i W_\psi x_i(a_j,b_i)\hat{\psi}_{a_j,b_i}(\omega) = \sum_i W_\psi\, \ddot{x}_g(a_j,b_i)H_i(\omega)\hat{\psi}_{a_j,b_i}(\omega) \tag{3.43}$$

$$\sum_i W_\psi x_c(a_j,b_i)\hat{\psi}_{a_j,b_i}(\omega) = \sum_i W_\psi\, \ddot{x}_g(a_j,b_i)H_c(\omega)\hat{\psi}_{a_j,b_i}(\omega) \tag{3.44}$$

Once the transfer function is obtained, the same procedure as described in Sections 2.5 and 2.6 may be followed to obtain the nonstationary peak responses of impulsive and convective modes of liquid vibration. However, it would be interesting to see the effects of ground motion nonstationarities on tank–liquid vibration. For this purpose, the hydrodynamic pressure on the tank wall is calculated based on wavelet coefficients. The expression of the hydrodynamic pressure on a tank wall due to vibration in the containing liquid caused by seismic ground motion is given below:

$$P_H(t) = \rho_l R\cos(\theta)\left[c_i(z)\ddot{x}_i(t) + c_c(R,z)\ddot{x}_c(t)\right] \tag{3.45}$$

In the expression of the hydrodynamic pressure, the terms $c_i(z)$ and $c_c(R,z)$ denote dimensionless functions defining the height-wise variation of hydrodynamic pressure and the distribution of the component of the hydrodynamic pressure corresponding to the first convective mode, respectively. These two distribution functions do not consider the rotational part of the vibration. The term θ in Equation 3.45 is assumed to be 0° because currently we are considering only the lateral component of seismic base excitation in our analysis. The expression for $c_c(R,z)$ in Equation (3.45) for hydrodynamic pressures on the wall of the tank is written as [27]

$$c_c(R,z) = \frac{2}{\lambda_1^2 - 1} \frac{\cosh\left(\lambda_1 \frac{z}{R}\right)}{\cosh\left(\lambda_1 \frac{H}{R}\right)} \tag{3.46}$$

As with the terms like $x_i(t)$ before, the term $P_H(t)$ representing the hydrodynamic pressure may also be expressed in terms of wavelet coefficients as

$$P_H(t) = \sum_i \sum_j \frac{K\Delta b}{a_j} W_\psi P_H(a_j, b_i) \psi_{a_j,b_i}(t) \tag{3.47}$$

Let us substitute Equations (3.31), (3.34) and (3.47) in Equation (3.45) to obtain the following:

$$\sum_i \sum_j \frac{K\Delta b}{a_j} W_\psi P_H(a_j, b_i)\psi_{a_j,b_i}(t) = \rho_l R\cos(\theta)$$
$$\left[c_i(z) \sum_i \sum_j \frac{K\Delta b}{a_j} W_\psi x_i(a_j, b_i) \ddot{\psi}_{a_j,b_i}(t) + c_c(R,z) \sum_i \sum_j \frac{K\Delta b}{a_j} W_\psi x_c(a_j, b_i) \ddot{\psi}_{a_j,b_i}(t) \right] \tag{3.48}$$

If we now take the Fourier transform of both the sides of Equation (3.48), multiply both sides by $\hat{\psi}^*_{a_k,b_l}(\omega)$ and use Equation (2.40), we get the following resulting equation:

$$\sum_i W_\psi P_H(a_j, b_i)\hat{\psi}_{a_j,b_i}(\omega) = -\omega^2 \rho_l R\cos(\theta)$$
$$\left[c_i(z) \sum_i W_\psi x_i(a_j, b_i)\hat{\psi}_{a_j,b_i}(\omega) + c_c(R,z) \sum_i W_\psi x_c(a_j, b_i)\hat{\psi}_{a_j,b_i}(\omega) \right] \tag{3.49}$$

On substituting the expressions for $\Sigma_i W_\psi x_i(a_j, b_i)\hat{\psi}_{a_j,b_i}(\omega)$ and $\Sigma_i W_\psi x_c(a_j, b_i)\hat{\psi}_{a_j,b_i}(\omega)$ from Equations (3.43) and (3.44) in Equation (3.49), one can easily find the following:

$$\sum_i W_\psi P_H(a_j, b_i)\hat{\psi}_{a_j,b_i}(\omega) = -\omega^2 \rho_l R\cos(\theta)$$
$$\left[c_i(z) \sum_i W_\psi \ddot{x}_g(a_j, b_i) H_i(\omega)\hat{\psi}_{a_j,b_i}(\omega) + c_c(R,z) \sum_i W_\psi \ddot{x}_g(a_j, b_i) H_c(\omega)\hat{\psi}_{a_j,b_i}(\omega) \right] \tag{3.50}$$

The above equation may be further simplified to

$$\sum_i W_\psi P_H(a_j,b_i)\hat{\psi}_{a_j,b_i}(\omega) = \sum_i W_\psi\, \ddot{x}_g(a_j,b_i)\tau_{PH}(\omega)\hat{\psi}_{a_j,b_i}(\omega) \tag{3.51}$$

Equation (3.50) expresses the relationship of hydrodynamic pressure to ground acceleration in terms of wavelet coefficients through a frequency-dependent transfer function, $\tau_{PH})(\omega)$. The transfer function $\tau_{PH}(\omega)$ has the following expression:

$$\tau_{PH}(\omega) = -\omega^2\rho_l R\cos(\theta)\left[c_i(z)H_i(\omega) + c_c(R,z)H_c(\omega)\right]\sum_i W_\psi\, \ddot{x}_g(a_j,b_i)\hat{\psi}_{a_j,b_i}(\omega) \tag{3.52}$$

The transfer function may be divided by maximum hydrostatic pressure at the tank bottom ($\rho_l Hg$) to give it a nondimensional form. Thus, on doing so and also using $\cos\theta = 1$ (as θ is assumed to be zero), the expression for the frequency-dependent nondimensional hydrodynamic pressure coefficient, $\tau_{CPH}(\omega)$, becomes

$$\tau_{CPH}(\omega) = -\frac{\omega^2}{g\frac{H}{R}}\left[c_i(z)H_i(\omega) + c_c(R,z)H_c(\omega)\right] \tag{3.53}$$

This hydrodynamic pressure coefficient may now be used to obtain the expected largest peak responses, as shown in Chapter 2.

3.5 NUMERICAL STUDY

The wavelet-based analytical approach described so far is now used to study the responses of a structural system. First, wavelet coefficients obtained from the data analysis of the time history record are demonstrated before and after ground motion characterization. Subsequently, a tank structure supported on the ground and assumed to be rigidly attached to the soil-foundation system is analysed by wavelets. The responses of the system are also obtained from exact time history analysis and comparisons are made between the two methods.

3.5.1 Ground motion characterization

The wavelet coefficients are calculated corresponding to the random ground acceleration time history representing the Loma Prieta earthquake (1989), as shown in Figure 1.2. The wavelet coefficients are obtained using

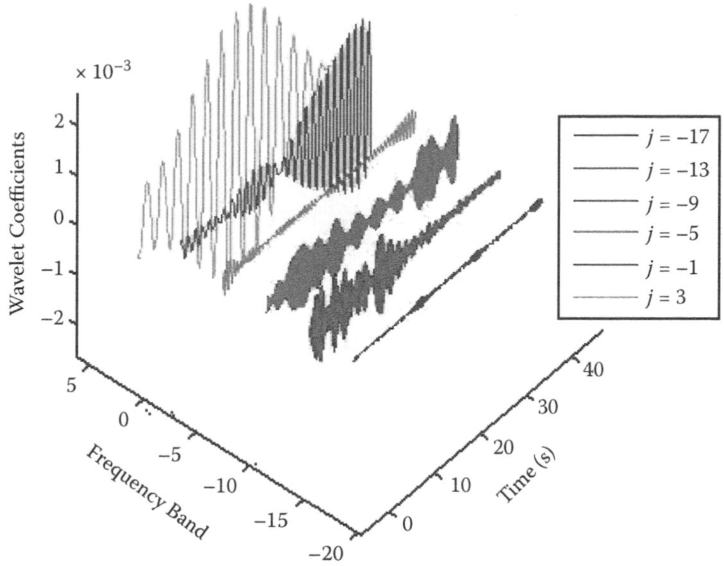

Figure 3.6 Wavelet coefficients of the Loma Prieta (1989) acceleration data for some frequency bands.

the modified Littlewood–Paley basis function as explained in Chapter 2 (Equation 2.36) for frequency bands and are depicted in Figure 3.6. The ground motion is then characterized as explained before by redistributing the wavelet coefficients at different time instants based on the central frequency of each frequency band, and then these coefficients are squared, as shown in Figure 3.7.

3.5.2 Validation – PSA response

Figures 3.8 and 3.9 demonstrate the curves representing pseudospectral accelerations (in terms of *g*) of a single-degree-of-freedom system for 1% damping and 5% damping, respectively. These figures reveal that the responses obtained from wavelet-based analysis of the dynamic SDOF system are in reasonably good agreement with the responses obtained from exact time history simulation (using the Runge–Kutta fourth-order method). It may be reiterated here that the wavelet-based approach has considered only a single time history of input seismic excitation, whereas the time history analysis is based upon an ensemble of several different accelerograms representing the input seismic excitation at the same site. With higher level of damping, the response gets less.

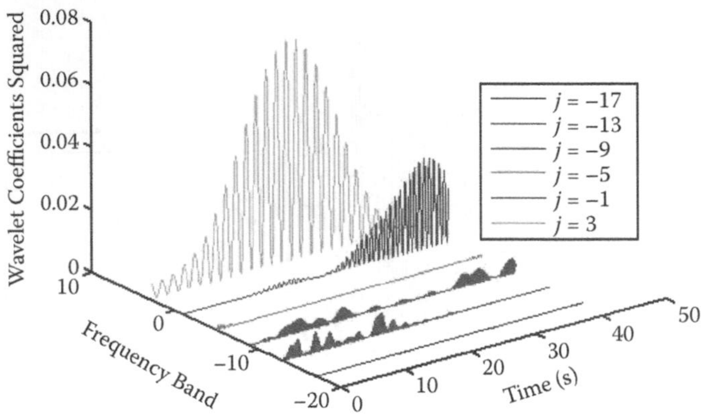

Figure 3.7 Squared wavelet coefficients after ground motion characterization for the Loma Prieta (1989) acceleration data for specific frequency bands.

3.5.3 Validation – structural response

A right circular, cylindrical water storage tank structure made of steel with a thick foundation is assumed to be rigidly attached to the ground surface. Hence, the rocking effects of vibration as well as any soil–structure interaction effects may be ignored. The tank is subjected to a random ground vibration. The tank contains water to its full height and has no top cover.

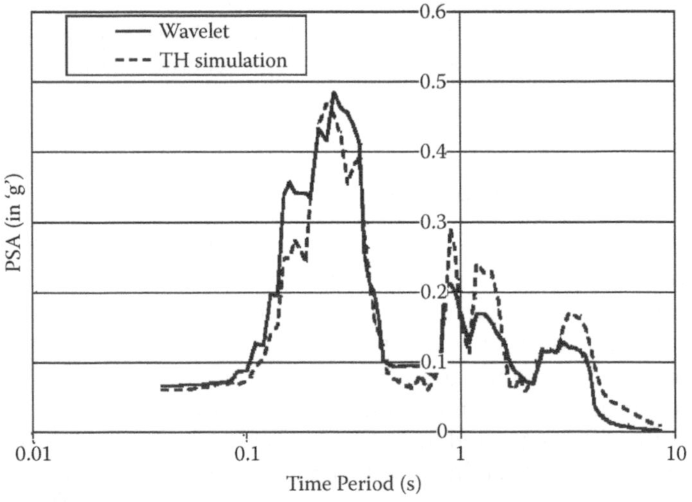

Figure 3.8 Comparison of PSA spectra obtained from time history simulation and wavelet analysis corresponding to 1% damping in the system.

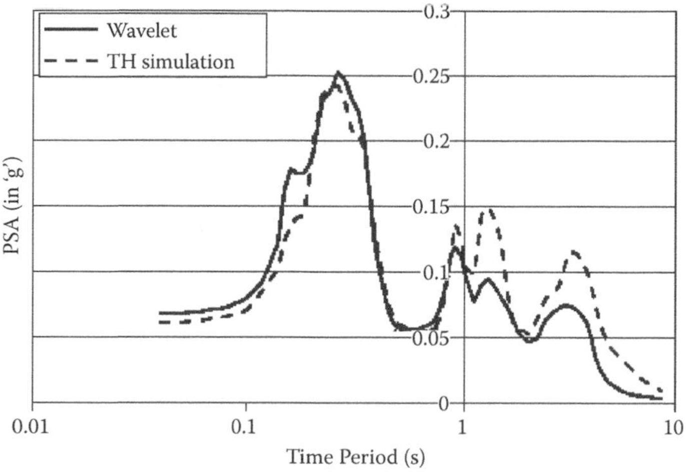

Figure 3.9 Comparison of PSA spectra obtained from time history simulation and wavelet analysis corresponding to 5% damping in the system.

The steel tank is assumed to have a damping ratio of 5%. The damping ratio ζ_i of the water mass vibrating in impulsive mode in unison with the tank is also assumed to be 5%. The damping ratio ζ_c corresponding to the convective mode of liquid vibration is considered to be 1%. The ratio of tank wall thickness to tank radius (i.e. $\frac{h}{R}$) is considered to be 0.001. The linear elastic material parameters for the steel tank are given by E (Young's modulus of elasticity) = 2.1×10^{11} N/m^2, ν (Poisson's ratio) = 0.3 and ρ (mass density) = 7850 kg/m^3. The ratio of the mass density of the tank water to the mass density of the steel tank is maintained constant at 0.127. The water in the tank is assumed to have a density of 1000 kg/m^3. A series of parametric variation has been done with respect to the height of the tank-liquid, H, for both fixed broad tanks with height-to-radius ratios less than or equal to 1 and for fixed slender tanks with height–radius ratios greater than unity. The hydrodynamic pressures and base shears are obtained at different heights along the tank wall and are plotted in graphs. Some of these are also compared with the results of the same system obtained using time history simulations. We will find the distribution of the hydrodynamic pressure coefficient on the tank wall along the vertical height. The height-radius ratio of the tank is kept fixed at 0.5 (i.e. tanks are chosen broader) and three different heights of the tank are considered: 3, 7 and 15 m. As per the analytical formulation, we need the values of λ_1, C_i, $C_i(z)$ and $C_c(z)$ to calculate the transfer function relating the pressure coefficients on the tank wall to the ground acceleration. As we are considering only the first sloshing mode of vibration of the liquid in the tank, the value of λ_1, which

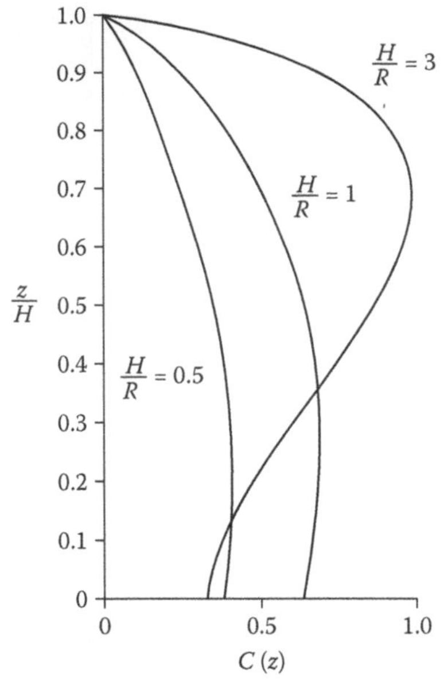

Figure 3.10 Variation of $C_i(z)$ on tank wall with different height–radius ratios of a tank. (From Veletsos, A. S., and Tang, Y., *Earthquake Eng. Struct. Dyn.*, 19, 473–496, 1990.)

is the first zero of the first derivative of the Bessel function [27], is 1.841. The values of C_i may be obtained from Table 3.1. The values of $C_i(z)$ representing impulsive components of hydrodynamic pressure can be obtained from Figure 3.10 [27], and the values of $C_c(z)$ are easily obtainable from Equation (3.46).

The hydrodynamic pressure coefficients are obtained from time history simulation and wavelet analysis for three different heights of the tank structure (H = 3, 7 and 10 m) and are shown in Figures 3.11 to 3.13.

In each figure, the two curves corresponding to two types of analysis are quite close to each other, demonstrating the reliability of the wavelet-based method of analysis. Thus, wavelet-based analysis based on a single representative accelerogram may be carried out to obtain the design curves for a given system. It is seen that the pressure increases with the increase in the height of the tank wall.

The coefficients of base shear are subsequently obtained for various tank heights for two different tank height–radius ratios (1.0 and 3.0). These are obtained from time history simulation and wavelet analysis and are shown in Figures 3.14 and 3.15. The curves seem to be in good agreement with each other. These validations give us confidence that the method adopted for ground motion characterization works well.

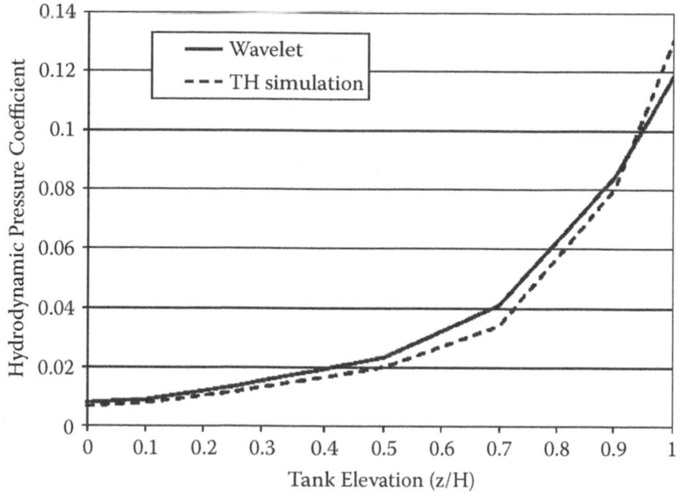

Figure 3.11 Hydrodynamic pressure coefficients on tank wall obtained from time history analysis and wavelet-based analysis for H = 3 m.

3.5.4 Wavelet analysis – structural response

For the benefit of the readers, some more wavelet analysis has been done with different height–radius ratios to generate the design curves in case of tanks fixed to the foundation. The other parameters remain the same, as

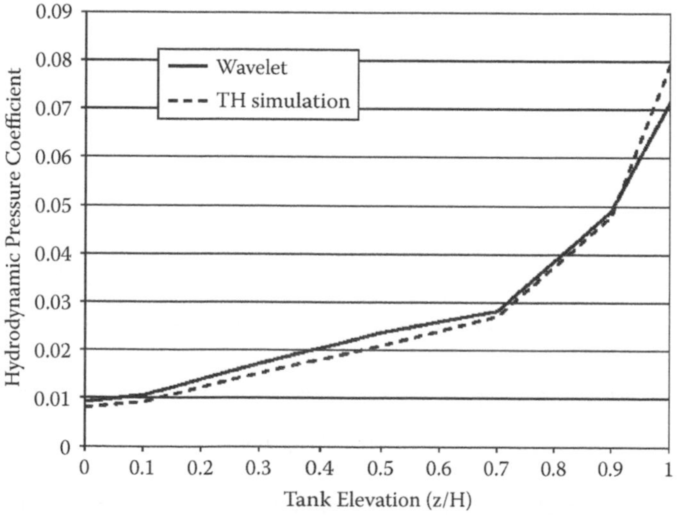

Figure 3.12 Hydrodynamic pressure coefficients on tank wall obtained from time history analysis and wavelet-based analysis for H = 7 m.

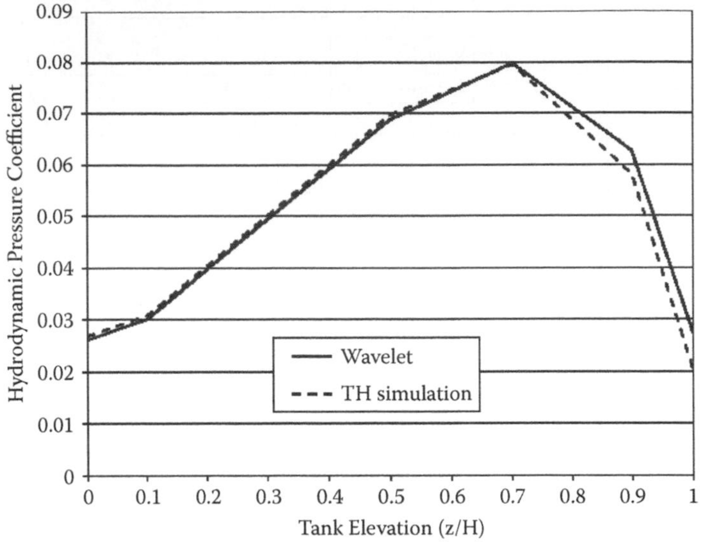

Figure 3.13 Hydrodynamic pressure coefficients on tank wall obtained from time history analysis and wavelet-based analysis for H = 15 m.

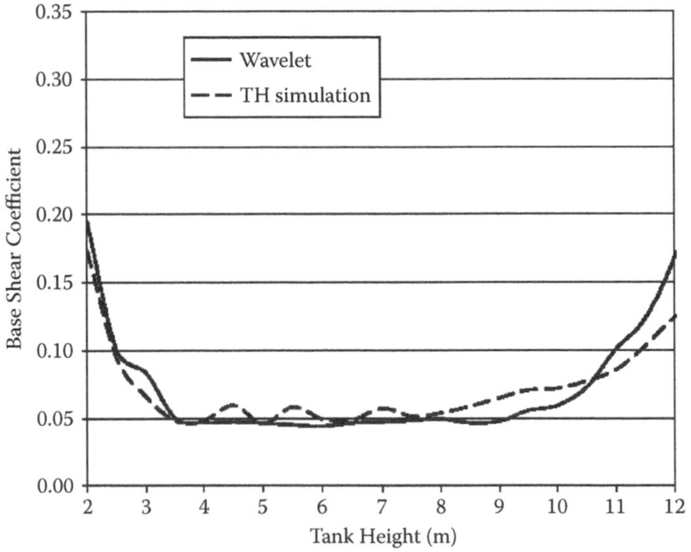

Figure 3.14 Variation of base shear coefficients with tank height obtained from time history analysis and wavelet-based analysis for tank height–radius ratio equal to 1.0.

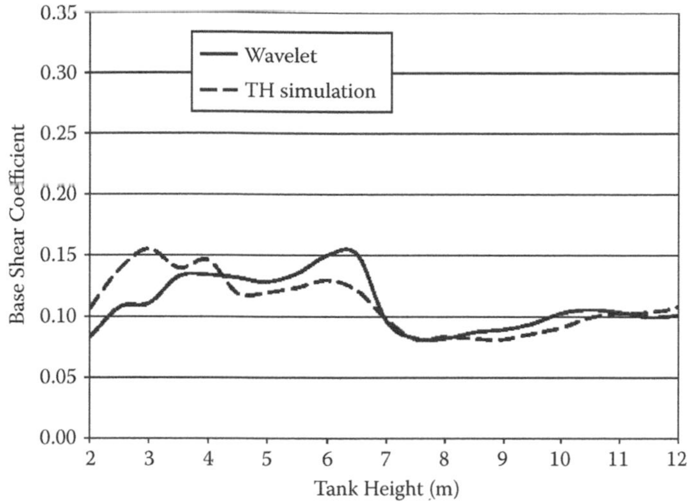

Figure 3.15 Variation of base shear coefficients with tank height obtained from time history analysis and wavelet-based analysis for tank height–radius ratio equal to 3.0.

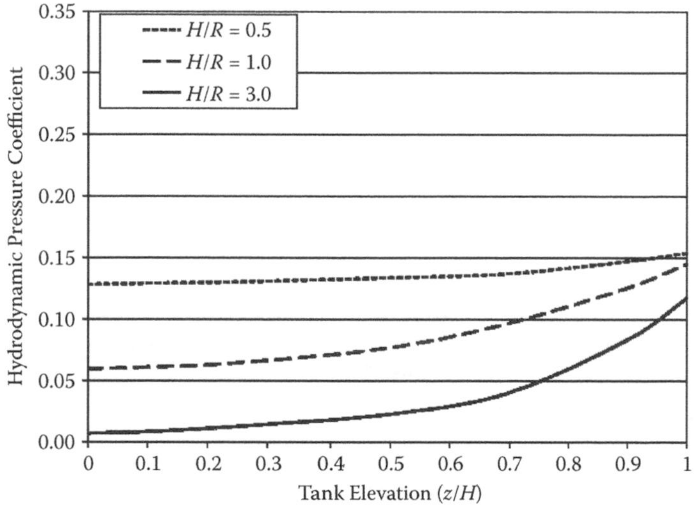

Figure 3.16 Pressure coefficients on tank wall obtained from wavelet analysis for H = 3 m for various height–radius ratios.

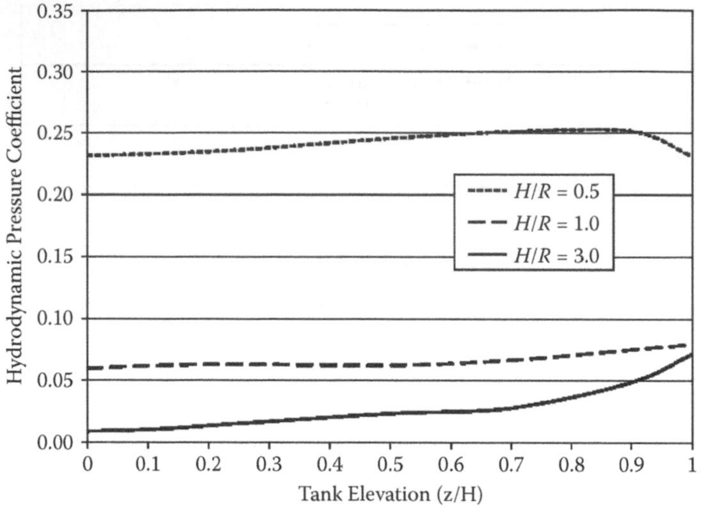

Figure 3.17 Pressure coefficients on tank wall obtained from wavelet analysis for H = 7 m for various height–radius ratios.

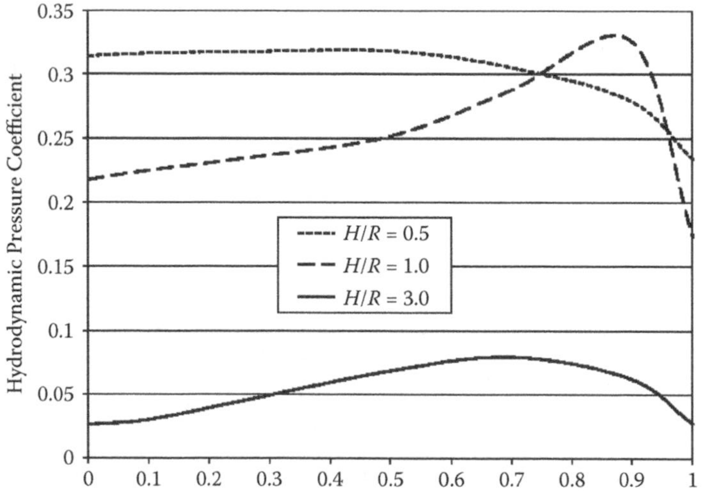

Figure 3.18 Pressure coefficients on tank wall obtained from wavelet analysis for H = 15 m for various height–radius ratios.

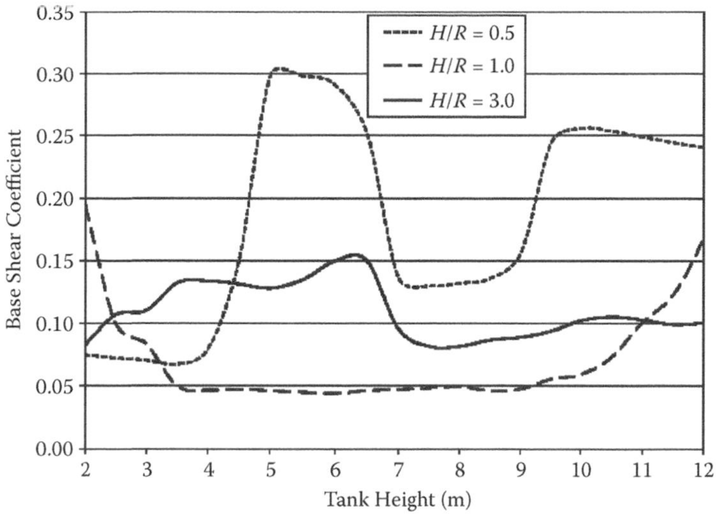

Figure 3.19 Shear coefficients at tank base obtained from wavelet analysis for different tank heights and height–radius ratios.

described in Section 3.5.3. The hydrodynamic pressure coefficients are calculated again for different height–radius ratios (0.5, 1.0 and 3.0) and are plotted in Figures 3.16 to 3.18 for tank heights 3, 7 and 15 m, respectively. Similarly, base shear coefficients for various tank heights are calculated for $H/R = 0.5$ using the wavelet-based technique and are shown in Figure 3.19 along with the wavelet-based analytical results already computed in the previous section for $H/R = 1.0$ and 3.0.

Chapter 4

Wavelet-based analysis of linear MDOF system

We have seen in Chapter 3 how wavelet-based formulation is gradually developed for single-degree-of-freedom (SDOF) and two-degree-of-freedom (2-DOF) systems. In case of the 2-DOF system, the equations representing the dynamic motion of the system were uncoupled. So, searching for a solution was not a difficult task. However, in reality most of the systems may not always be representable as a simplified SDOF or 2-DOF system; rather, they would need to be represented with more degrees of freedom. In this chapter, we will take up a unique example of a specific multi-degree-of-freedom (MDOF) system and show step-wise how to make valid assumptions to concentrate more on those parameters that would affect the system response to a great extent. The first important consideration would be to include soil interaction in the analysis. In Chapter 3 we have assumed the superstructure to be anchored to rigid foundations, thereby ignoring soil–structure interaction. The other important criterion could be consideration of rocking motion of the structure [29, 30]. During strong ground shakings, it is quite likely that ground-supported structures like liquid storage tanks, buildings, reactors, etc. would be subjected to rocking motions as well. With these considerations, the main objective of this chapter is to define a suitable model with practical assumptions, development of a theoretical formulation based on the wavelet-based technique and providing a closed-form analytical solution to the problem. The basic guidelines for such formulation are well proposed in the work by Chatterjee and Basu [31].

4.1 DESCRIPTION OF THE MODEL

The ground-supported cylindrical liquid storage tank is idealized as a multi-degree-of-freedom (in this case, 4-DOF) system. The tank is assumed to be flexibly attached to the soil–foundation system. The depth of the liquid is the same as the height of the tank. The liquid in the tank is assumed to vibrate in predominantly two modes, as explained in Chapter 3. One is impulsive mode of vibration, in which the liquid (water in this case) moves

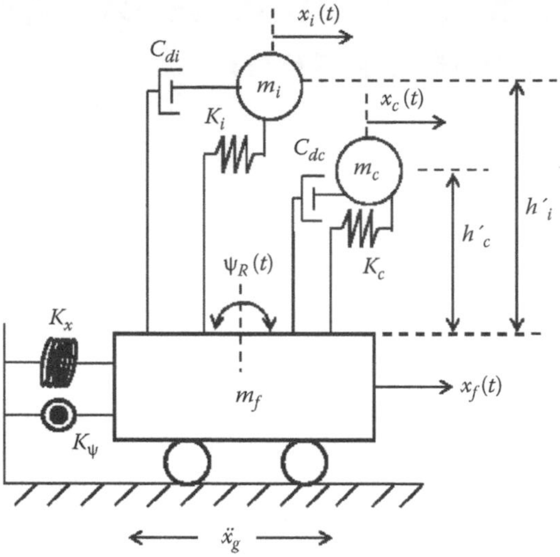

Figure 4.1 MDOF model for a tank–fluid–foundation–soil system.

in unison with the tank. The other is the convective mode of liquid vibration, in which the water body vibrates mostly in the first sloshing mode. The tank is resting on a circular footing (of same radius as that of the tank), which is assumed to undergo rocking motion. Thus, the equivalent tank–liquid–foundation–soil system consists of 4 degrees of freedom – the horizontal displacement of impulsive liquid mass, the horizontal displacement of convective liquid mass, the horizontal displacement of the tank–foundation and the rocking motion of the tank–foundation system. The tank–liquid–foundation–soil model subjected to a seismic base excitation is shown in Figure 4.1.

The two modes of liquid vibration are represented by two different masses – impulsive mass m_i and convective mass m_c lumped at heights h'_i and h'_c, respectively, above the base of the tank. The impulsive mass of the tank–liquid is assumed to be attached to the tank through a combined spring-damper system, which actually represents the stiffness and corresponding damping in the impulsive mode. The spring coefficient and the viscous damping coefficient are denoted by K_i and C_{di}, respectively. Similar terms for the convective liquid mass are denoted by K_c and C_{dc}, respectively. The soil–foundation system is represented by the foundation mass m_f, which is attached to the surrounding soil medium through two sets of spring-dashpot combinations. These two sets of spring-dashpot systems are represented by the terms K_x and K_ψ, respectively – the first one being the complex-valued translational impedance function for the circular footing on which the tanks rests, and the second term denotes the same for rocking motion of that footing.

The model is subjected to horizontal free-field ground displacement, $x_g(t)$. The ground shaking induces vibrations in the foundation, and hence it undergoes both lateral and rocking motions. The terms $x_f(t)$ and $\psi_R(t)$ denote the absolute lateral and rocking displacements of the foundation. The displacements of the impulsive and convective masses with respect to the foundation are denoted by $x_i(t)$ and $x_c(t)$, respectively. The linear static stiffness coefficients, K_i and K_c for impulsive and convective liquid masses, respectively, are defined as

$$K_i = m_i\omega_i^2 \tag{4.1}$$

$$K_c = m_c\omega_c^2 \tag{4.2}$$

The linear viscous damping coefficients, C_{di} and C_{dc}, are defined as

$$C_{di} = 2m_i\zeta_i\omega_i \tag{4.3}$$

$$C_{dc} = 2m_c\zeta_c\omega_c \tag{4.4}$$

Now, the reader must be willing to know the expressions for natural frequencies for impulsive and convective (first sloshing) modes of liquid vibration, i.e. ω_i and ω_c, respectively. These expressions are given below [2]:

$$\omega_i = \frac{C_i}{H}\sqrt{\frac{E}{\rho}} \tag{4.5}$$

$$\omega_c = 2\pi\sqrt{\lambda_1 \tanh\left(\lambda_1 \frac{H}{R}\right)\frac{g}{R}} \tag{4.6}$$

In Equation (4.5), C_i is a dimensionless coefficient and is a function of four parameters. The first one is the ratio of liquid height to tank radius (for the present case, it is H/R), and the second one is the ratio of tank wall thickness (h) to tank radius (R). The next parameter is the Poisson's ratio of the tank material (denoted by ν). The other parameter is the relative mass density of the liquid with respect to that of the tank (i.e. $\frac{\rho_l}{\rho}$). The terms E and ρ denote the linear elastic tank material parameters, Young's modulus and mass density, respectively. In Equation (4.6), λ_1 is the first root of the first derivative of the Bessel function of the first kind and the first order and g is the acceleration due to gravity.

The convective mass (m_c) of the liquid forms a part of the total liquid mass in the tank and is calculated from the height–radius ratio of the tank (λ_1) and the total mass of the tank–liquid (m_1). The expression for m_c is given as

$$m_c = \left[\frac{2}{\lambda_1^2}\left(\frac{R}{\lambda_1 H}\right)\tanh\left(\lambda_1 \frac{H}{R}\right)\right] m_l \tag{4.7}$$

The height of the lumped convective mass (h_c') above the tank base is given as follows:

$$h_c' = h_c + \Delta h_c \tag{4.8}$$

where

$$h_c = \left[1 - \frac{1}{\lambda_1}\frac{R}{H}\tanh\left(\frac{\lambda_1}{2}\frac{H}{R}\right)\right] H \tag{4.9}$$

and

$$\Delta h_c = \frac{H}{\lambda_1 \dfrac{H}{R}\sinh\left(\lambda_1 \dfrac{H}{R}\right)} \tag{4.10}$$

4.2 EQUATIONS OF MOTION

Now let us formulate the equations of dynamic equilibrium for the given system. As rocking motion is considered, the readers may note that the liquid masses would undergo additional translation due to rocking besides normal translational displacement due to horizontal motion. Let us first consider the equilibrium of only the impulsive mass of the liquid. The corresponding equation of motion may be written as

$$m_i[\ddot{x}_i(t) + \ddot{x}_f(t)] + m_i h_i' \ddot{\psi}_R(t) + K_i[x_i(t) + h_i' \psi_R(t)] + C_{di}[\dot{x}_i(t) + h_i' \dot{\psi}_R(t)] = 0 \tag{4.11}$$

On using Equations (4.1) and (4.3), Equation (4.11) may be simplified to the following:

$$m_i[\ddot{x}_i(t)+\ddot{x}_f(t)]+m_i h_i' \ddot{\psi}_R(t)+m_i\omega_i^2[x_i(t)+h_i' \psi_R(t)]$$
$$+2\zeta_i m_i\omega_i[\dot{x}_i(t)+h_i' \dot{\psi}_R(t)]=0 \qquad (4.12)$$

The single and double dots represent single and double differentiations with respect to time. On considering the equilibrium of only the convective mass of the liquid and using relations given by Equations (4.2) and (4.4), one may write

$$m_c[\ddot{x}_c(t)+\ddot{x}_f(t)]+m_c h_c' \ddot{\psi}_R(t)+m_c\omega_c^2[x_c(t)+h_c'\psi_R(t)]$$
$$+2\zeta_c m_c\omega_c[\dot{x}_c(t)+h_c'\dot{\psi}_R(t)]=0 \qquad (4.13)$$

Next, we consider the equilibrium of the tank–liquid–foundation system. This gives us the third equation of dynamic motion as follows:

$$m_i[\ddot{x}_i(t)+\ddot{x}_f(t)]+m_c[\ddot{x}_c(t)+\ddot{x}_f(t)]+m_i h_i' \ddot{\psi}_R(t)+m_c h_c' \ddot{\psi}_R(t)+m_f\ddot{x}_f(t)+S_B=0 \qquad (4.14)$$

The term S_B in the above equation denotes the shear at the tank base (precisely at the interface between the base of the tank–foundation and the supporting soil). Assuming that the foundation (i.e. the circular footing on which the tank is mounted) does not slip over the soil surface, the base shear remains proportional to the relative deformation of the tank–foundation system and the soil. Thus, the base shear may be expressed as

$$S_B = K_x(x_f(t)-x_g(t)) \qquad (4.15)$$

The term K_x, as said earlier, represents the complex-valued translational impedance function of the circular footing in consideration and has the following expression:

$$K_x = \frac{8G_sR}{2-\nu_s}(\alpha_x + ia_0\beta_x) \qquad (4.16)$$

The term G_s represents the shear modulus of the soil medium, which is related to the mass density of the soil (ρ) and the shear wave velocity in the soil (V_s) as per the following relation:

$$G_s = \rho V_s^2 \qquad (4.17)$$

There are three other terms in Equation (4.16): a_0, α_x and β_x. The first term, a_0, is a nondimensional frequency parameter related to the frequency (ω) of foundation vibration through the following relation:

$$a_0 = \frac{\omega R}{V_s} \tag{4.18}$$

The other terms, α_x and β_x, are the dimensionless functions of a_0 and ν_s. Veletsos and Verbic [32], in one of their papers, proposed closed-form solutions for these two functions. It may be noted here that for the static case (i.e. if the value of $\omega = 0$), the imaginary component $ia_0\beta_x$ vanishes in Equation (4.16), and subsequently the term K_x consists of only real terms, which represents the static stiffness coefficient.

The fourth and last equation may be developed based on the equilibrium of moments of the tank–liquid–foundation system and written as

$$\frac{1}{2} m_f R^2 \ddot{\psi}_R(t) + K_\psi \psi_R(t) + m_i h_i' \left[\ddot{x}_i(t) + \ddot{x}_f(t) + h_i' \ddot{\psi}_R(t)\right]$$

$$+ m_c h_c' \left[\ddot{x}_c(t) + \ddot{x}_f(t) + h_c' \ddot{\psi}_R(t)\right] = 0 \tag{4.19}$$

In Equation (4.19), the term K_ψ denotes the linear rocking impedance function of the foundation, which depends upon the soil properties and the frequency of foundation vibration. The expression for K_ψ [33] is given below:

$$K_\psi = \frac{8 G_s R^3}{3(1 - \nu_s)} (\alpha_\psi + i a_0 \beta_\psi) \tag{4.20}$$

The terms α_ψ and β_ψ are the two dimensionless functions that depend on the nondimensional frequency parameter a_0, as defined by Equation (4.18) as follows:

$$\alpha_\psi = 1 - \frac{0.3178 a_0^2}{1 + 0.675 a_0^2} \tag{4.21}$$

$$\beta_\psi = \frac{0.2612 a_0^2}{1 + 0.675 a_0^2} \tag{4.22}$$

It may be mentioned here that the above expressions of α_ψ and β_ψ are valid corresponding to the value of $\nu_s = \frac{1}{3}$.

Furthermore, let us transform Equations (4.16) and (4.20) to obtain the expressions for lateral and rocking natural frequencies of vibration of the tank–liquid–foundation system. Let us divide both sides of these two equations by m_f and $\frac{1}{2}m_f R^2$, respectively, as follows:

$$\frac{K_x}{m_f} = \frac{8G_s R}{m_f(2-\nu_s)}(\alpha_x + ia_0\beta_x) = \omega_{lf}^2(\alpha_x + ia_0\beta_x) \tag{4.23}$$

$$\frac{K_\psi}{\frac{1}{2}m_f R^2} = \frac{16G_s R}{3m_f(1-\nu_s)}(\alpha_\psi + ia_0\beta_\psi) = \omega_{rf}^2(\alpha_\psi + ia_0\beta_\psi) \tag{4.24}$$

The constant terms in front of the parenthesis on the right-hand sides of Equations (4.23) and (4.24) represent the squares of the natural frequencies for lateral (ω_{lf}^2) and rocking (ω_{rf}^2) motions of the foundation. These terms have the following expressions;

$$\omega_{lf} = \sqrt{\frac{8G_s R}{m_f(2-\nu_s)}} \tag{4.25}$$

$$\omega_{rf} = \sqrt{\frac{16G_s R}{3m_f(1-\nu_s)}} \tag{4.26}$$

4.3 WAVELET DOMAIN FORMULATION OF TANK–LIQUID–FOUNDATION SYSTEM

The dynamic equilibrium of the tank–liquid–foundation–soil system is governed by four equations, (4.12) to (4.14) and (4.19), as shown in the preceding section. The MDOF system is subjected to a seismic ground acceleration process with a given nonstationary statistical characteristic. In this section we will make an attempt to transform the equations to the wavelet domain. The wavelet domain-coupled dynamic equations are formulated and then solved to get the expressions of instantaneous power spectral density function (PSDF) in terms of the functionals of input wavelet coefficients. The moments of the instantaneous PSDF are used to obtain the stochastic responses of the tank in the form of coefficients of hydrodynamic pressure, base shear and overturning base moment for the largest expected peak responses. A few parametric variations are carried out to study the effects of various governing parameters, e.g. height of liquid in the tank, height–radius ratio of the tank, ratio of total liquid mass to mass of foundation, shear wave velocity in the soil medium, etc.

As seen in Chapter 2, like any other nonstationary system response, the nonstationary ground displacement and ground acceleration may also be expressed in terms of wavelet coefficients as

$$x_g(t) = \sum_i \sum_j \frac{K\Delta b}{a_j} W_\psi x_g(a_j, b_i) \psi_{a_j,b_i}(t) \tag{4.27}$$

$$\ddot{x}_g(t) = \sum_i \sum_j \frac{K\Delta b}{a_j} W_\psi \ddot{x}_g(a_j, b_i) \psi_{a_j,b_i}(t) \tag{4.28}$$

On differentiating Equation (4.27) two times and subsequently taking the Fourier transform of both sides, one may get

$$\hat{\ddot{x}}_g(\omega) = \sum_i \sum_j \frac{K\Delta b}{a_j} W_\psi x_g(a_j, b_i) \psi_{a_j,b_i}(\omega)(-\omega^2) \tag{4.29}$$

On taking the Fourier transform of both sides of Equation (4.28), one may also get

$$\hat{\ddot{x}}_g(\omega) = \sum_i \sum_j \frac{K\Delta b}{a_j} W_\psi \ddot{x}_g(a_j, b_i) \psi_{a_j,b_i}(\omega) \tag{4.30}$$

On comparing Equations (4.29) and (4.30), one may obtain the following important relation, which we will make use of:

$$\sum_i \sum_j \frac{K\Delta b}{a_j} W_\psi x_g(a_j, b_i) \hat{\psi}_{a_j,b_i}(\omega) = \left(-\frac{1}{\omega^2}\right) \sum_i \sum_j \frac{K\Delta b}{a_j} W_\psi \ \ddot{x}_g(a_j, b_i) \hat{\psi}_{a_j,b_i}(\omega) \tag{4.31}$$

The above equation relates the wavelet coefficients of ground displacement to the wavelet coefficients of ground acceleration. As also explained in Chapter 2, the absolute lateral and rocking displacements of the foundation may be expressed in terms of wavelet coefficients as follows:

$$x_f(t) = \sum_i \sum_j \frac{K\Delta b}{a_j} W_\psi x_f(a_j, b_i) \psi_{a_j,b_i}(t) \tag{4.32}$$

$$\psi_R(t) = \sum_i \sum_j \frac{K\Delta b}{a_j} W_\psi \psi_R(a_j, b_i) \psi_{a_j,b_i}(t) \tag{4.33}$$

Similarly, the system responses in terms of relative displacements of impulsive and convective liquid masses with respect to the tank–foundation may be expanded in terms of wavelet coefficients as follows:

$$x_i(t) = \sum_i \sum_j \frac{K\Delta b}{a_j} W_\psi x_i(a_j, b_i) \psi_{a_j,b_i}(t) \tag{4.34}$$

$$x_c(t) = \sum_i \sum_j \frac{K\Delta b}{a_j} W_\psi x_c(a_j, b_i) \psi_{a_j,b_i}(t) \tag{4.35}$$

The time derivatives of the above terms (single derivative indicating velocity and double derivative indicating acceleration) may be written as follows:

$$\dot{x}_f(t) = \sum_i \sum_j \frac{K\Delta b}{a_j} W_\psi x_f(a_j, b_i) \dot{\psi}_{a_j,b_i}(t) \tag{4.36}$$

$$\ddot{x}_f(t) = \sum_i \sum_j \frac{K\Delta b}{a_j} W_\psi x_f(a_j, b_i) \ddot{\psi}_{a_j,b_i}(t) \tag{4.37}$$

$$\dot{\psi}_R(t) = \sum_i \sum_j \frac{K\Delta b}{a_j} W_\psi \psi_R(a_j, b_i) \dot{\psi}_{a_j,b_i}(t) \tag{4.38}$$

$$\ddot{\psi}_R(t) = \sum_i \sum_j \frac{K\Delta b}{a_j} W_\psi \psi_R(a_j, b_i) \ddot{\psi}_{a_j,b_i}(t) \tag{4.39}$$

$$\dot{x}_i(t) = \sum_i \sum_j \frac{K\Delta b}{a_j} W_\psi x_i(a_j, b_i) \dot{\psi}_{a_j,b_i}(t) \tag{4.40}$$

$$\ddot{x}_i(t) = \sum_i \sum_j \frac{K\Delta b}{a_j} W_\psi x_i(a_j, b_i) \dot{\psi}_{a_j,b_i}(t) \tag{4.41}$$

$$\dot{x}_c(t) = \sum_i \sum_j \frac{K\Delta b}{a_j} W_\psi x_c(a_j, b_i) \dot{\psi}_{a_j,b_i}(t) \tag{4.42}$$

$$\ddot{x}_c(t) = \sum_i \sum_j \frac{K\Delta b}{a_j} W_\psi x_c(a_j, b_i) \ddot{\psi}_{a_j,b_i}(t) \tag{4.43}$$

On substituting the expressions for $\ddot{x}_i(t)$, $\ddot{x}_f(t)$, $\ddot{\psi}_R(t)$, $x_i(t)$, $\psi_R(t)$, $\dot{x}_i(t)$ and $\dot{\psi}_R(t)$ in Equation (4.12) and cancelling the common term, m_i, one may get

$$\begin{aligned}
&\sum_i \sum_j \frac{K\Delta b}{a_j} W_\psi x_i(a_j, b_i) \ddot{\psi}_{a_j,b_i}(t) + \sum_i \sum_j \frac{K\Delta b}{a_j} W_\psi x_f(a_j, b_i) \ddot{\psi}_{a_j,b_i}(t) + \\
&h_i' \sum_i \sum_j \frac{K\Delta b}{a_j} W_\psi \psi_R(a_j, b_i) \ddot{\psi}_{a_j,b_i}(t) + \\
&\omega_i^2 \Bigg[\sum_i \sum_j \frac{K\Delta b}{a_j} W_\psi x_i(a_j, b_i) \psi_{a_j,b_i}(t) + \\
&\quad + h_i' \sum_i \sum_j \frac{K\Delta b}{a_j} W_\psi \psi_R(a_j, b_i) \psi_{a_j,b_i}(t) \Bigg] \\
&\quad + 2\zeta_i \omega_i \Bigg[\sum_i \sum_j \frac{K\Delta b}{a_j} W_\psi x_i(a_j, b_i) \dot{\psi}_{a_j,b_i}(t) \\
&+ h_i' \sum_i \sum_j \frac{K\Delta b}{a_j} W_\psi \psi_R(a_j, b_i) \dot{\psi}_{a_j,b_i}(t) \Bigg] = 0
\end{aligned} \tag{4.44}$$

On taking the Fourier transform of both sides, Equation (4.44) becomes

$$\begin{aligned}
&\sum_i \sum_j \frac{K\Delta b}{a_j} W_\psi x_i(a_j, b_i) \hat{\psi}_{a_j,b_i}(\omega)(-\omega^2) + \sum_i \sum_j \frac{K\Delta b}{a_j} W_\psi x_f(a_j, b_i) \hat{\psi}_{a_j,b_i}(\omega)(-\omega^2) + \\
&h_i' \sum_i \sum_j \frac{K\Delta b}{a_j} W_\psi \psi_R(a_j, b_i) \hat{\psi}_{a_j,b_i}(\omega)(-\omega^2) + \omega_i^2 \\
&\Bigg[\sum_i \sum_j \frac{K\Delta b}{a_j} W_\psi x_i(a_j, b_i) \hat{\psi}_{a_j,b_i}(\omega) + h_i' \sum_i \sum_j \frac{K\Delta b}{a_j} W_\psi \psi_R(a_j, b_i) \hat{\psi}_{a_j,b_i}(\omega) \Bigg] \\
&+ 2\zeta_i \omega_i \Bigg[\sum_i \sum_j \frac{K\Delta b}{a_j} W_\psi x_i(a_j, b_i) \hat{\psi}_{a_j,b_i}(\omega)(i\omega) + \\
&h_i' \sum_i \sum_j \frac{K\Delta b}{a_j} W_\psi \psi_R(a_j, b_i) \hat{\psi}_{a_j,b_i}(\omega)(i\omega) \Bigg] = 0
\end{aligned} \tag{4.45}$$

On collecting similar terms, Equation (4.45) reduces to

$$\left[\omega_i^2 - \omega^2 + 2i\zeta_i\omega_i\omega\right]\sum_i\sum_j \frac{K\Delta b}{a_j} W_\psi x_i(a_j,b_i)\hat{\psi}_{a_j,b_i}(\omega) +$$

$$[-\omega^2]\sum_i\sum_j \frac{K\Delta b}{a_j} W_\psi x_f(a_j,b_i)\hat{\psi}_{a_j,b_i}(\omega) + \tag{4.46}$$

$$\left[\omega_i^2 h_i' - h_i'\omega^2 + 2i\zeta_i\omega_i\omega h_i'\right]\sum_i\sum_j \frac{K\Delta b}{a_j} W_\psi \psi_R(a_j,b_i)\hat{\psi}_{a_j,b_i}(\omega) = 0$$

Now, multiplying both sides of Equation (4.46) by $\hat{\psi}^*_{a_k,b_l}(\omega)$ and using Equation (2.40) of Chapter 2, the following equation is obtained:

$$\left[\omega_i^2 - \omega^2 + 2i\zeta_i\omega_i\omega\right]\sum_i W_\psi x_i(a_j,b_i)\,\hat{\psi}_{a_j,b_i}(\omega) + [-\omega^2]\sum_i W_\psi x_f(a_j,b_i)\,\hat{\psi}_{a_j,b_i}(\omega) +$$

$$h_i'\left[\omega_i^2 - \omega^2 + 2i\zeta_i\omega_i\omega\right]\sum_i W_\psi \psi_R(a_j,b_i)\hat{\psi}_{a_j,b_i}(\omega) = 0 \tag{4.47}$$

Now, let us take up the next equation of system equilibrium, which is Equation (4.13). On substituting the expressions for $\ddot{x}_c(t)$, $\ddot{x}_f(t)$, $\ddot{\psi}_R(t)$, $x_c(t)$, $\psi_R(t)$, $\dot{x}_c(t)$ and $\dot{\psi}_R(t)$ in Equation (4.13) and cancelling the common term, m_c, one may get

$$\sum_i\sum_j \frac{K\Delta b}{a_j} W_\psi x_c(a_j,b_i)\ddot{\psi}_{a_j,b_i}(t) + \sum_i\sum_j \frac{K\Delta b}{a_j} W_\psi x_f(a_j,b_i)\ddot{\psi}_{a_j,b_i}(t) +$$

$$h_c'\sum_i\sum_j \frac{K\Delta b}{a_j} W_\psi \psi_R(a_j,b_i)\ddot{\psi}_{a_j,b_i}(t) +$$

$$\omega_c^2\left[\sum_i\sum_j \frac{K\Delta b}{a_j} W_\psi x_c(a_j,b_i)\psi_{a_j,b_i}(t) + h_c'\sum_i\sum_j \frac{K\Delta b}{a_j} W_\psi \psi_R(a_j,b_i)\psi_{a_j,b_i}(t)\right] +$$

$$2\zeta_c\omega_c\left[\sum_i\sum_i \frac{K\Delta b}{a_j} W_\psi x_c(a_j,b_i)\dot{\psi}_{a_j,b_i}(t) +\right.$$

$$\left. h_c'\sum_i\sum_j \frac{K\Delta b}{a_j} W_\psi \psi_R(a_j,b_i)\dot{\psi}_{a_j,b_i}(t)\right] = 0 \tag{4.48}$$

On taking the Fourier transform of both sides, Equation (4.48) becomes

$$\sum_i\sum_j \frac{K\Delta b}{a_j} W_\psi x_c(a_j,b_i)\hat{\psi}_{a_j,b_i}(\omega)(-\omega^2)+\sum_i\sum_j \frac{K\Delta b}{a_j} W_\psi x_f(a_j,b_i)\hat{\psi}_{a_j,b_i}(\omega)(-\omega^2)$$

$$+h_c'\sum_i\sum_j \frac{K\Delta b}{a_j} W_\psi \psi_R(a_j,b_i)\hat{\psi}_{a_j,b_i}(\omega)(-\omega^2)+$$

$$\omega_c^2\left[\sum_i\sum_j \frac{K\Delta b}{a_j} W_\psi x_c(a_j,b_i)\hat{\psi}_{a_j,b_i}(\omega)+h_c'\sum_i\sum_j \frac{K\Delta b}{a_j} W_\psi \psi_R(a_j,b_i)\hat{\psi}_{a_j,b_i}(\omega)\right]$$

$$+2\zeta_c\omega_c\left[\sum_i\sum_j \frac{K\Delta b}{a_j} W_\psi x_c(a_j,b_i)\hat{\psi}_{a_j,b_i}(\omega)(i\omega)+\right.$$

$$\left.h_C'\sum_i\sum_j \frac{K\Delta b}{a_j} W_\psi \psi_R(a_j,b_i)\hat{\psi}_{a_j,b_i}(\omega)(i\omega)\right]=0 \tag{4.49}$$

On collecting similar terms, Equation (4.49) takes the following form:

$$\sum_i\sum_j \frac{K\Delta b}{a_j} W_\psi x_c(a_j,b_i)\hat{\psi}_{a_j,b_i}(\omega)\left[\omega_c^2-\omega^2+2i\zeta_C\omega_c\omega\right]+$$

$$[-\omega^2]\sum_i\sum_j \frac{K\Delta b}{a_j} W_\psi x_f(a_j,b_i)\hat{\psi}_{a_j,b_i}(\omega)+$$

$$\left[\omega_c^2 h_c'-h_c'\omega^2+2i\zeta_c\omega_c\omega h_c'\right]\sum_i\sum_j \frac{K\Delta b}{a_j} W_\psi \psi_R(a_j,b_i)\hat{\psi}_{a_j,b_i}(\omega)=0 \tag{4.50}$$

Now, we shall use the same technique here. First, we multiply both sides of Equation (4.50) by $\hat{\psi}^*_{a_k,b_l}(\omega)$ and use Equation (2.40) to obtain the following equation:

$$\left[\omega_c^2-\omega^2+2i\zeta_c\omega_c\omega\right]\sum_i W_\psi x_c(a_j,b_i)\hat{\psi}_{a_j,b_i}(\omega)+[-\omega^2]\sum_i W_\psi x_f(a_j,b_i)\hat{\psi}_{a_j,b_i}(\omega)$$

$$+h_c'\left[\omega_c^2-\omega^2+2i\zeta_c\omega_c\omega\right]\sum_i W_\psi \psi_R(a_j,b_i)\hat{\psi}_{a_j,b_i}(\omega)=0 \tag{4.51}$$

Now, we pick up the third equation pertaining to the dynamic equilibrium of the tank–liquid–foundation system, which is Equation (4.14). On replacing the base shear, S_B, by its expression as given in Equation (4.15), on substituting the expressions for $x_f(t)$, $x_g(t)$, $\ddot{x}_i(t)$, $\ddot{x}_c(t)$, $\ddot{x}_f(t)$ and $\ddot{\psi}_R(t)$ in Equation (4.14) and dividing throughout by m_f, one can get the following equation:

$$\frac{m_i}{m_f}\sum_i\sum_j \frac{K\Delta b}{a_j} W_\psi x_i(a_j,b_i)\ddot{\psi}_{a_j,b_i}(t) + \left(\frac{m_i}{m_f}+\frac{m_c}{m_f}+1\right)$$

$$\sum_i\sum_j \frac{K\Delta b}{a_j} W_\psi x_f(a_j,b_i)\ddot{\psi}_{a_j,b_i}(t) +$$

$$\frac{m_c}{m_f}\sum_i\sum_j \frac{K\Delta b}{a_j} W_\psi x_c(a_j,b_i)\ddot{\psi}_{a_j,b_i}(t) + \left(h_i'\frac{m_i}{m_f}+h_c'\frac{m_c}{m_f}\right)$$

$$\sum_i\sum_j \frac{K\Delta b}{a_j} W_\psi \psi_R(a_j,b_i)\ddot{\psi}_{a_j,b_i}(t) + \frac{K_x}{m_f}\sum_i\sum_j \frac{K\Delta b}{a_j} W_\psi x_f(a_j,b_i)\psi_{a_j,b_i}(t) - \frac{K_x}{m_f}$$

$$\sum_i\sum_j \frac{K\Delta b}{a_j} W_\psi x_g(a_j,b_i)\psi_{a_j,b_i}(t) = 0 \tag{4.52}$$

We take the Fourier transform on both sides to transform it to the frequency domain as follows:

$$\frac{m_i}{m_f}\sum_i\sum_j \frac{K\Delta b}{a_j} W_\psi x_i(a_j,b_i)\hat{\psi}_{a_j,b_i}(\omega)(-\omega^2) +$$

$$\left(\frac{m_i}{m_f}+\frac{m_c}{m_f}+1\right)\sum_i\sum_j \frac{K\Delta b}{a_j} W_\psi x_f(a_j,b_i)\ddot{\psi}_{a_j,b_i}(t)\hat{\psi}_{a_j,b_i}(\omega)(-\omega^2) +$$

$$\frac{m_c}{m_f}\sum_i\sum_j \frac{K\Delta b}{a_j} W_\psi x_c(a_j,b_i)\hat{\psi}_{a_j,b_i}(\omega)(-\omega^2) +$$

$$\left(h_i'\frac{m_i}{m_f}+h_c'\frac{m_c}{m_f}\right)\sum_i\sum_j \frac{K\Delta b}{a_j} W_\psi \psi_R(a_j,b_i)\hat{\psi}_{a_j,b_i}(\omega)(-\omega^2) +$$

$$\frac{K_x}{m_f}\sum_i\sum_j \frac{K\Delta b}{a_j} W_\psi x_f(a_j,b_i)\hat{\psi}_{a_j,b_i}(\omega) - \frac{K_x}{m_f}\sum_i\sum_j \frac{K\Delta b}{a_j} W_\psi x_g(a_j,b_i)\hat{\psi}_{a_j,b_i}(\omega) = 0$$

$$\tag{4.53}$$

Now, again we multiply both sides of Equation (4.53) by $\hat{\psi}^*_{a_k,b_l}(\omega)$, use Equation (2.40) and divide all terms by ω^2 to obtain the following equation:

$$\frac{m_i}{m_f}\sum_i W_\psi x_i(a_j,b_i)\hat{\psi}_{a_j,b_i}(\omega)+\left(\frac{m_i}{m_f}+\frac{m_c}{m_f}+1-\frac{K_x}{\omega^2 m_f}\right)\sum_i W_\psi x_f(a_j,b_i)\hat{\psi}_{a_j,b_i}(\omega)$$
$$+\frac{m_c}{m_f}\sum_i W_\psi x_c(a_j,b_i)\hat{\psi}_{a_j,b_i}(\omega)+\left(h_i'\frac{m_i}{m_f}+h_c'\frac{m_c}{m_f}\right)\sum_i W_\psi \psi_R(a_j,b_i)\hat{\psi}_{a_j,b_i}(\omega)$$
$$=\frac{K_x}{\omega^4 m_f}\sum_i W_\psi \ddot{x}_g(a_j,b_i)\hat{\psi}_{a_j,b_i}(\omega)$$
(4.54)

We now take up the remaining system equilibrium equation (Equation 4.19) to convert it to the wavelet domain. For this we repeat the same process as done above with three other equations of dynamic equilibrium of the system. First, we substitute the expressions for $\ddot{x}_i(t)$, $\ddot{x}_c(t)$, $\ddot{x}_f(t)$, $\psi_R(t)$ and $\ddot{\psi}_R(t)$ in Equation (4.19) and divide throughout by m_f to obtain the following equation in the wavelet domain:

$$\frac{1}{2}R^2\sum_i\sum_j\frac{K\Delta b}{a_j}W_\psi\psi_R(a_j,b_i)\ddot{\psi}_{a_j,b_i}(t)+K_\psi\sum_i\sum_j\frac{K\Delta b}{a_j}W_\psi\psi_R(a_j,b_i)\psi_{a_j,b_i}(t)$$
$$+m_ih_i'\left[\sum_i\sum_j\frac{K\Delta b}{a_j}W_\psi x_i(a_j,b_i)\ddot{\psi}_{a_j,b_i}(t)+\sum_i\sum_j\frac{K\Delta b}{a_j}W_\psi x_f(a_j,b_i)\ddot{\psi}_{a_j,b_i}(t)\right.$$
$$\left.+h_i'\sum_i\sum_j\frac{K\Delta b}{a_j}W_\psi\psi_R(a_j,b_i)\ddot{\psi}_{a_j,b_i}(t)\right]$$
$$+m_ch_c'\left[\sum_i\sum_j\frac{K\Delta b}{a_j}W_\psi x_c(a_j,b_i)\ddot{\psi}_{a_j,b_i}(t)+\sum_i\sum_j\frac{K\Delta b}{a_j}W_\psi x_f(a_j,b_i)\ddot{\psi}_{a_j,b_i}(t)\right.$$
$$\left.+h_c'\sum_i\sum_j\frac{K\Delta b}{a_j}W_\psi\psi_R(a_j,b_i)\ddot{\psi}_{a_j,b_i}(t)\right]=0 \qquad (4.55)$$

Subsequently, we Fourier transform the wavelet domain equation to get the following:

$$
\begin{aligned}
&\frac{1}{2}R^2\sum_i\sum_j\frac{K\Delta b}{a_j}W_\psi\psi_R(a_j,b_i)\hat{\psi}_{a_j,b_i}(\omega)(-\omega^2)\\
&+K_\psi\sum_i\sum_j\frac{K\Delta b}{a_j}W_\psi\psi_R(a_j,b_i)\ddot{\psi}_{a_j,b_i}(\omega)\\
&+m_ih_i'\left[\sum_i\sum_j\frac{K\Delta b}{a_j}W_\psi x_i(a_j,b_i)\hat{\psi}_{a_j,b_i}(\omega)(-\omega^2)\right.\\
&+\sum_i\sum_j\frac{K\Delta b}{a_j}W_\psi x_f(a_j,b_i)\hat{\psi}_{a_j,b_i}(\omega)(-\omega^2)\\
&\left.+h_i'\sum_i\sum_j\frac{K\Delta b}{a_j}W_\psi\psi_R(a_j,b_i)\hat{\psi}_{a_j,b_i}(\omega)(-\omega^2)\right]\\
&+m_ch_c'\left[\sum_i\sum_j\frac{K\Delta b}{a_j}W_\psi x_c(a_j,b_i)\hat{\psi}_{a_j,b_i}(\omega)(-\omega^2)\right.\\
&+\sum_i\sum_j\frac{K\Delta b}{a_j}W_\psi x_f(a_j,b_i)\hat{\psi}_{a_j,b_i}(\omega)(-\omega^2)\\
&\left.+h_c'\sum_i\sum_j\frac{K\Delta b}{a_j}W_\psi\psi_R(a_j,b_i)\hat{\psi}_{a_j,b_i}(\omega)(-\omega^2)=0\right]
\end{aligned}
\tag{4.56}
$$

Now, we multiply both sides of Equation (4.56) by $\hat{\psi}^*_{a_k,b_l}(\omega)$, use Equation (2.40), divide all terms by ω^2 and then collect similar terms to obtain the following equation:

$$
\begin{aligned}
&\left(-\frac{m_i}{m_f}h_i'\right)\sum_i W_\psi x_i(a_j,b_i)\hat{\psi}_{a_j,b_i}(\omega)+\left(-\frac{m_i}{m_f}h_i'-\frac{m_c}{m_c}h_c\right)\sum_i W_\psi x_f(a_j,b_i)\hat{\psi}_{a_j,b_i}(\omega)\\
&+\left(-\frac{m_c}{m_f}h_c'\right)\sum_i W_\psi x_c(a_j,b_i)\hat{\psi}_{a_j,b_i}(\omega)+\left(-\frac{1}{2}R^2-\frac{m_i}{m_f}h_i'^2-\frac{m_c}{m_f}h_c'^2+\frac{K_\psi}{m_{f\omega^2}}\right)\\
&\times\sum_i W_\psi\psi_R(a_j,b_i)\hat{\psi}_{a_j,b_i}(\omega)=0
\end{aligned}
\tag{4.57}
$$

Thus, we obtain four coupled equations based on which we shall obtain the various responses of the tank–liquid–foundation-soil system in the wavelet domain.

4.4 WAVELET-BASED NONSTATIONARY SYSTEM RESPONSES

Let us temporarily denote the expressions $\sum_i W_\psi x_i(a_j,b_i)\hat{\psi}_{a_j,b_i}(\omega)$, $\sum_i W_\psi x_f(a_j,b_i)\hat{\psi}_{a_j,b_i}(\omega)$, $\sum_i W_\psi x_c(a_j,b_i)\hat{\psi}_{a_j,b_i}(\omega)$, $\sum_i W_\psi \psi_R(a_j,b_i)\hat{\psi}_{a_j,b_i}(\omega)$ and $\sum_i W_\psi \ddot{x}_g(a_j,b_i)\hat{\psi}_{a_j,b_i}(\omega)$ by X^i, X^c, X^f, X^ψ and $\ddot{X}^g$ for easier recognition in the remaining part of the formulation. So, Equations (4.47), (4.51), (4.54) and (4.57) may be respectively rewritten as

$$\left[\omega_i^2-\omega^2+2i\zeta_i\omega_i\omega\right]X^i+[-\omega^2]X^f+h_i'\left[\omega_i^2-\omega^2+2i\zeta_i\omega_i\omega\right]X^\psi=0 \quad (4.58)$$

$$\left[\omega_c^2-\omega^2+2i\zeta_c\omega_c\omega\right]X^c+[-\omega^2]X^f+h_c'\left[\omega_c^2-\omega^2+2i\zeta_c\omega_c\omega\right]X^\psi=0 \quad (4.59)$$

$$\frac{m_i}{m_f}X^i+\left(\frac{m_i}{m_f}+\frac{m_c}{m_f}+1-\frac{K_x}{\omega^2 m_f}\right)X^f+\frac{m_c}{m_f}X^c+\left(h_i'\frac{m_i}{m_f}+h_c'\frac{m_c}{m_f}\right)X^\psi=\frac{K_x}{\omega^4 m_f}\ddot{X}^g \quad (4.60)$$

$$\left(-\frac{m_i}{m_f}h_i'\right)X^i+\left(-\frac{m_i}{m_f}h_i'-\frac{m_c}{m_f}h_c'\right)X^f+\left(-\frac{m_c}{m_f}h_c'\right)X^c+\left(-\frac{1}{2}R^2-\frac{m_i}{m_f}h_i'^2-\frac{m_c}{m_f}h_c'^2+\frac{K_\psi}{\omega^2 m_f}\right)X^\psi=0 \quad (4.61)$$

However, our main objective in this chapter is to formulate the wavelet-based response of the tank in terms of hydrodynamic pressure on the tank wall, base shear and overturning base moment. The total hydrodynamic pressure, $P_H(t)$, at any point on the tank wall at a certain time instant is obtained by summing up the hydrodynamic pressures due to the impulsive component and the convective component (corresponding to the first sloshing mode only) at that time instant. Thus, the expression for $P_H(t)$ becomes

$$P_H(t)=\rho_l R\cos(\theta)[c_i(z)\{\ddot{x}_i(t)+\ddot{x}_f(t)+h_i'\ddot{\psi}_R(t)\}+c_c(R,z)\{\ddot{x}_c(t)+\ddot{x}_f(t)+h_c'\ddot{\psi}_R(t)\}] \quad (4.62)$$

We see two dimensionless functions used in the expression of hydrodynamic pressure. These are $c_i(z)$ and $c_c(R,z)$. The first function ($c_i(z)$)

represents the height-wise variation of hydrodynamic pressure for the impulsive component of liquid vibration, and the second one ($c_c(R,z)$) defines the distribution of the same due to the first convective (sloshing) mode. It is worth mentioning here that these two distribution functions do not consider the rotational component of the motion; however, the rocking component of the ground motion is included in Equation (4.62). It may also be noted that we are considering the seismic base excitation in the horizontal direction. Hence, the value of θ in Equation (4.62) may safely be assumed to be zero. So, the term $R\cos(\theta)$ becomes only R. The expression for $c_c(R,z)$ is given below [27], which may be used to obtain the hydrodynamic pressure on the tank wall (at $r = R$).

$$c_c(R,z) = \frac{2}{\lambda_1^2 - 1} \frac{\cosh\left(\lambda_1 \frac{z}{R}\right)}{\cosh\left(\lambda_1 \frac{H}{R}\right)} \tag{4.63}$$

The effective shear at the tank base (Q_B) and the net effective overturning moment (M_B) at any time at a section immediately above the tank base may be calculated from Figure 4.1 in terms of impulsive displacement, convective displacement, rocking response and the foundation displacement and are written below:

$$Q_B(t) = m_i\{\ddot{x}_i(t) + \ddot{x}_f(t) + h_i' \ddot{\psi}_R(t)\} + m_c\{\ddot{x}_C(t) + \ddot{x}_f(t) + h_c' \ddot{\psi}_R(t)\} \tag{4.64}$$

$$M_B(t) = m_i h_i\{\ddot{x}_i(t) + \ddot{x}_f(t) + h_i' \ddot{\psi}_R(t)\} + m_c h_c\{\ddot{x}_C(t) + \ddot{x}_f(t) + h_c \ddot{\psi}_R(t)\} \tag{4.65}$$

As we have done earlier, the next step would be to represent the hydrodynamic pressure, base shear and base moment in terms of wavelet coefficients as follows:

$$P_H(t) = \sum_i \sum_j \frac{K\Delta b}{a_j} W_\psi P_H(a_j, b_i) \psi_{a_j,b_i}(t) \tag{4.66}$$

$$Q_B(t) = \sum_i \sum_j \frac{K\Delta b}{a_j} W_\psi Q_B(a_j, b_i) \psi_{a_j,b_i}(t) \tag{4.67}$$

$$M_B(t) = \sum_i \sum_j \frac{K\Delta b}{a_j} W_\psi M_B(a_j, b_i) \psi_{a_j,b_i}(t) \tag{4.68}$$

On substituting the expressions for $\ddot{x}_f(t)$, $\ddot{\psi}_R(t)$, $\ddot{x}_i(t)$, $\ddot{x}_c(t)$ and $P_H(t)$ from Equations (4.37), (4.39), (4.41), (4.43) and (4.66) in Equation (4.62), one may obtain the following expression for hydrodynamic pressure in the wavelet domain.

$$\begin{aligned}
&\sum_i\sum_j \frac{K\Delta b}{a_j} W_\psi P_H(a_j,b_i)\psi_{a_j,b_i}(t) \\
&= \rho_l R\left[c_i(z)\left\{\sum_i\sum_j \frac{K\Delta b}{a_j} W_\psi x_i(a_j,b_i)\ddot{\psi}_{a_j,b_i}(t) + \sum_i\sum_j \frac{K\Delta b}{a_j} W_\psi x_f(a_j,b_i)\ddot{\psi}_{a_j,b_i}(t) \right.\right.\\
&\left. + h_i' \sum_i\sum_j \frac{K\Delta b}{a_j} W_\psi \psi_R(a_j,b_i)\ddot{\psi}_{a_j,b_i}(t)\right\} \\
&+ c_c(R,z)\left\{\sum_i\sum_j \frac{K\Delta b}{a_j} W_\psi x_c(a_j,b_i)\ddot{\psi}_{a_j,b_i}(t) + \sum_i\sum_j \frac{K\Delta b}{a_j} W_\psi x_f(a_j,b_i)\ddot{\psi}_{a_j,b_i}(t) \right.\\
&\left.\left. + h_c' \sum_i\sum_j \frac{K\Delta b}{a_j} W_\psi \psi_R(a_j,b_i)\ddot{\psi}_{a_j,b_i}(t)\right\}\right]
\end{aligned} \tag{4.69}$$

Now, on Fourier transforming Equation (4.69), multiplying both sides of the transformed equation by $\hat{\psi}^*_{a_k,b_l}(\omega)$ and subsequently using Equation (2.40), one may obtain the following expression of hydrodynamic pressure in terms of wavelet functionals:

$$\sum_i W_\psi P_H(a_j,b_i)\hat{\psi}_{a_j,b_i}(\omega) = \tau_H(\omega)\sum_i W_\psi \ddot{x}_g(a_j,b_i)\hat{\psi}_{a_j,b_i}(\omega) \tag{4.70}$$

Equation (4.70) relates the wavelet coefficients of hydrodynamic pressure on the inner tank wall at various heights to the wavelet coefficients of the ground acceleration through the transfer function $\tau_{PH}(\omega)$, which has the following expression:

$$\begin{aligned}
\tau_H(\omega) = -\rho_l R\ \omega^2[c_i(z)\{\tau_i(\omega) + \tau_f(\omega) + h_i'\tau_R(\omega)\} + c_c(R,z)\{\tau_c(\omega) + \tau_f(\omega) \\
+ h_c'\tau_R(\omega)\}]
\end{aligned} \tag{4.71}$$

This transfer function for hydrodynamic pressure is not like the other transfer functions we have seen before. The previous transfer functions were relating the wavelet coefficients of a specific response, say, impulsive displacement, directly with those of the ground acceleration using system properties and excitation frequency. In this case, $\tau_H(\omega)$ involves several other transfer functions to establish the relation between wavelet coefficients of hydrodynamic pressure and wavelet coefficients of ground motion. In Equation (4.71), $\tau_i(\omega)$, $\tau_c(\omega)$ and $\tau_f(\omega)$ represent the transfer functions relating the displacements of the impulsive mode of the liquid, the first convective mode of the liquid and the foundation to the ground acceleration. The term $\tau_R(\omega)$ represents the transfer function relating the rocking motion of the foundation to the ground acceleration. All these relationships are valid in terms of wavelet coefficients. At this point, let us assume the expressions for $\tau_i(\omega)$, $\tau_c(\omega)$, $\tau_f(\omega)$ and $\tau_R(\omega)$, as mentioned below:

$$\sum_i W_\psi x_i(a_j,b_i)\hat{\psi}_{a_j,b_i}(\omega)=\tau_i(\omega)\sum_i W_\psi\ \ddot{x}_g(a_j,b_i)\hat{\psi}_{a_j,b_i}(\omega) \tag{4.72}$$

$$\sum_i W_\psi x_c(a_j,b_i)\hat{\psi}_{a_j,b_i}(\omega)=\tau_c(\omega)\sum_i W_\psi \ddot{x}_g(a_j,b_i)\hat{\psi}_{a_j,b_i}(\omega) \tag{4.73}$$

$$\sum_i W_\psi x_f(a_j,b_i)\hat{\psi}_{a_j,b_i}(\omega)=\tau_f(\omega)\sum_i W_\psi\ \ddot{x}_g(a_j,b_i)\hat{\psi}_{a_j,b_i}(\omega) \tag{4.74}$$

$$\sum_i W_\psi \psi_R(a_j,b_i)\hat{\psi}_{a_j,b_i}(\omega)=\tau_R(\omega)\sum_i W_\psi\ \ddot{x}_g(a_j,b_i)\hat{\psi}_{a_j,b_i}(\omega) \tag{4.75}$$

In fact, the above expressions may even be simplified if we use the naming convention as described at the beginning of Section 4.4 and rewritten as

$$X^i=\tau_i(\omega)\ \ddot{X}^g \tag{4.76}$$

$$X^c=\tau_c(\omega)\ \ddot{X}^g \tag{4.77}$$

$$X^f=\tau_f(\omega)\ \ddot{X}^g \tag{4.78}$$

$$X^\psi=\tau_R(\omega)\ \ddot{X}^g \tag{4.79}$$

On substituting the expressions for $\ddot{x}_f(t)$, $\ddot{\psi}_R(t)$, $\ddot{x}_i(t)$, $\ddot{x}_c(t)$ and $Q_B(t)$ from Equations (4.37), (4.39), (4.41), (4.43) and (4.67) in Equation (4.64), one may obtain the following expression for base shear in the wavelet domain:

$$\begin{aligned}
&\sum_i\sum_j \frac{K\Delta b}{a_j} W_\psi Q_B(a_j,b_i)\psi_{a_j,b_i}(t)\\
&=m_i\left[\sum_i\sum_j \frac{K\Delta b}{a_j} W_\psi x_i(a_j,b_i)\ddot{\psi}_{a_j,b_i}(t)+\sum_i\sum_j \frac{K\Delta b}{a_j} W_\psi x_f(a_j,b_i)\ddot{\psi}_{a_j,b_i}(t)\right.\\
&\left.+h_i'\sum_i\sum_j \frac{K\Delta b}{a_j} W_\psi \psi_R(a_j,b_i)\ddot{\psi}_{a_j,b_i}(t)\right]+m_c\left[\left\{\sum_i\sum_j \frac{K\Delta b}{a_j} W_\psi x_c(a_j,b_i)\ddot{\psi}_{a_j,b_i}(t)\right.\right.\\
&\left.\left.+\sum_i\sum_j \frac{K\Delta b}{a_j} W_\psi x_f(a_j,b_i)\ddot{\psi}_{a_j,b_i}(t)+h_c'\sum_i\sum_j \frac{K\Delta b}{a_j} W_\psi \psi_R(a_j,b_i)\ddot{\psi}_{a_j,b_i}(t)\right]\right\}
\end{aligned} \tag{4.80}$$

As done earlier, taking the Fourier transform of Equation (4.80), multiplying both sides of the transformed equation by $\hat{\psi}^{*}_{a_k,b_l}(\omega)$ and then using Equation (2.40), the relation between the wavelet coefficients of tank base shear and the wavelet coefficients of ground acceleration may be obtained through an appropriate transfer function, $\tau_Q(\omega)$, as follows:

$$\sum_i W_\psi Q_B(a_j,b_i)\hat{\psi}_{a_j,b_i}(\omega) = \tau_Q(\omega)\sum_i W_\psi \ddot{x}_g(a_j,b_i)\hat{\psi}_{a_j,b_i}(\omega) \tag{4.81}$$

The expression for $\tau_Q(\omega)$ is shown below:

$$\tau_Q(\omega) = -\omega^2[m_i\{\tau_i(\omega)+\tau_f(\omega)+h_i'\tau_R(\omega)\}+m_c\{\tau_c(\omega)+\tau_f(\omega)+h_c'\tau_R(\omega)\}] \tag{4.82}$$

Similarly, on substituting the expressions for $\ddot{x}_f(t)$, $\ddot{\psi}_R(t)$, $\ddot{x}_i(t)$, $\ddot{x}_c(t)$ and $M_B(t)$ from Equations (4.37), (4.39), (4.41), (4.43) and (4.68) in Equation (4.64), one may obtain the following expression for the overturning base moment in the wavelet domain:

$$\begin{aligned}
&\sum_i\sum_j \frac{K\Delta b}{a_j} W_\psi M_B(a_j,b_i)\psi_{a_j,b_i}(t)\\
&= m_ih_i\left[\sum_i\sum_j \frac{K\Delta b}{a_j} W_\psi x_i(a_j,b_i)\ddot{\psi}_{a_j,b_i}(t) + \sum_i\sum_j \frac{K\Delta b}{a_j} W_\psi x_f(a_j,b_i)\ddot{\psi}_{a_j,b_i}(t)\right.\\
&\left. + h_i' \sum_i\sum_j \frac{K\Delta b}{a_j} W_\psi \psi_R(a_j,b_i)\ddot{\psi}_{a_j,b_i}(t)\right]\\
&+ m_ch_c\left[\left\{\sum_i\sum_j \frac{K\Delta b}{a_j} W_\psi x_c(a_j,b_i)\ddot{\psi}_{a_j,b_i}(t)\right.\right.\\
&\left.\left. + \sum_i\sum_j \frac{K\Delta b}{a_j} W_\psi x_f(a_j,b_i)\ddot{\psi}_{a_j,b_i}(t) + h_c' \sum_i\sum_j \frac{K\Delta b}{a_j} W_\psi \psi_R(a_j,b_i)\ddot{\psi}_{a_j,b_i}(t)\right]\right\}
\end{aligned} \tag{4.83}$$

Following a similar sequence of operations as done above in the cases of hydrodynamic pressure and base shear, the following expression for base moment may be obtained:

$$\sum_i W_\psi Q_B(a_j,b_i)\hat{\psi}_{a_j,b_i}(\omega) = \tau_M(\omega)\sum_i W_\psi \ddot{x}_g(a_j,b_i)\hat{\psi}_{a_j,b_i}(\omega) \tag{4.84}$$

in which the transfer function $\tau_M(\omega)$ has the following expression:

$$\tau_M(\omega) = -\omega^2[m_i h_i\{\tau_i(\omega)+\tau_f(\omega)+h_i'\tau_R(\omega)\}+m_c h_c\{\tau_c(\omega)+\tau_f(\omega)+h_c'\tau_R(\omega)\}] \tag{4.85}$$

Now, it is clear to us that we have to know beforehand the expressions for $\tau_i(\omega)$, $\tau_c(\omega)$, $\tau_f(\omega)$ and $\tau_R(\omega)$ to obtain the transfer functions $\tau_H(\omega)$, $\tau_Q(\omega)$ and $\tau_M(\omega)$. This is possible if we solve Equations (4.58) to (4.61) to get the expressions for the unknowns X^i, X^c, X^f and X^ψ in terms of the known variable, $\ddot{X}^g$.

4.5 SOLUTION OF TRANSFER FUNCTIONS

Equations (4.58) to (4.61) form a set of simple linear simultaneous algebraic equations that may be readily expressed in the following matrix format:

$$[C_{coeff}]\begin{bmatrix} X^i \\ X^c \\ X^f \\ X^\psi \end{bmatrix} = \begin{bmatrix} 0 \\ 0 \\ \dfrac{K_x}{\omega^4 m_f} \\ 0 \end{bmatrix}[\ddot{X}^g] \tag{4.86}$$

The complex coefficients are placed in matrix $[C_{coeff}]$ given by

$[C_{coeff}]=$

$$\begin{bmatrix} \omega_i^2-\omega^2+2i\zeta_i\omega_i\omega & 0 & -\omega^2 & h_i'\{\omega_i^2-\omega^2+2i\zeta_i\omega_i\omega\} \\ 0 & \omega_c^2-\omega^2+2i\zeta_c\omega_c\omega & -\omega^2 & h_c'\{\omega_c^2-\omega^2+2i\zeta_c\omega_c\omega\} \\ \dfrac{m_i}{m_f} & \dfrac{m_c}{m_f} & \dfrac{m_i}{m_f}+\dfrac{m_c}{m_f}+1-\dfrac{K_x}{\omega^2 m_f} & \left\{h_i'\dfrac{m_i}{m_f}+h_c'\dfrac{m_c}{m_f}\right\} \\ -\dfrac{m_i}{m_f}h_i' & -\dfrac{m_c}{m_f}h_c' & -\dfrac{m_i}{m_f}h_i'-\dfrac{m_c}{m_f}h_c' & \left\{-\dfrac{1}{2}R^2-\dfrac{m_i}{m_f}h_i'^2-\dfrac{m_c}{m_f}h_c'^2+\dfrac{K_\psi}{\omega^2 m_f}\right\} \end{bmatrix} \tag{4.87}$$

Now, the solutions for X^i, X^c, X^f and X^ψ may be obtained from Equations (4.86) once we are able to invert the complex coefficient matrix, $[C_{coeff}]$. Let us assume that on inverting matrix $[C_{coeff}]$, we obtain real and imaginary components as follows:

$$[C_{coeff}]^{-1} = [C_{coeff}]^R + i\,[C_{coeff}]^i \tag{4.88}$$

where

$$[C_{coeff}]^R = \begin{bmatrix} c_{11}^r & c_{12}^r & c_{13}^r & c_{14}^r \\ c_{21}^r & c_{22}^r & c_{23}^r & c_{24}^r \\ c_{31}^r & c_{32}^r & c_{33}^r & c_{34}^r \\ c_{41}^r & c_{42}^r & c_{43}^r & c_{44}^r \end{bmatrix} \tag{4.89}$$

and

$$[C_{coeff}]^i = \begin{bmatrix} c_{11}^i & c_{12}^i & c_{13}^i & c_{14}^i \\ c_{21}^i & c_{22}^i & c_{23}^i & c_{24}^i \\ c_{31}^i & c_{32}^i & c_{33}^i & c_{34}^i \\ c_{41}^i & c_{42}^i & c_{43}^i & c_{44}^i \end{bmatrix} \tag{4.90}$$

Thus, multiplying both sides of Equation (4.5.1) by $[C_{coeff}]^{-1}$, one may obtain

$$\begin{bmatrix} X^i \\ X^c \\ X^f \\ X^\Psi \end{bmatrix} = \left[\left\{ \begin{matrix} c_{11}^r & c_{12}^r & c_{13}^r & c_{14}^r \\ c_{21}^r & c_{22}^r & c_{23}^r & c_{24}^r \\ c_{31}^r & c_{32}^r & c_{33}^r & c_{34}^r \\ c_{41}^r & c_{42}^r & c_{43}^r & c_{44}^r \end{matrix} \right\} + i \left\{ \begin{matrix} c_{11}^i & c_{12}^i & c_{13}^i & c_{14}^i \\ c_{21}^i & c_{22}^i & c_{23}^i & c_{24}^i \\ c_{31}^i & c_{32}^i & c_{33}^i & c_{34}^i \\ c_{41}^i & c_{42}^i & c_{43}^i & c_{44}^i \end{matrix} \right\}\right]$$

$$\begin{bmatrix} 0 \\ 0 \\ \dfrac{K_x}{\omega^4 m_f} \\ 0 \end{bmatrix} [\ddot{X}^g] \tag{4.91}$$

From Equation (4.91), the expression for X^i (i.e. $\sum_i W_\psi x_i(a_j,b_i)\hat{\psi}_{a_j,b_i}(\omega)$) may be derived in terms of $\ddot{X}^g$ (i.e. $\sum_i W_\psi\ \ddot{x}_g(a_j,b_i)\hat{\psi}_{a_j,b_i}(\omega)$) as

$$X^i = \frac{K_x}{\omega^4 m_f}\left\{c_{13}^r + ic_{13}^i\right\}[\ddot{X}^g] \tag{4.92}$$

Similarly, other relations between wavelet coefficients of convective displacement, foundation displacement and base rotation may be written in terms of wavelet coefficients of seismic ground acceleration as

$$X^c = \frac{K_x}{\omega^4 m_f}\left\{c_{23}^r + ic_{23}^i\right\}[\ddot{X}^g] \tag{4.93}$$

$$X^f = \frac{K_x}{\omega^4 m_f}\left\{c_{33}^r + ic_{33}^i\right\}[\ddot{X}^g] \tag{4.94}$$

$$X^\psi = \frac{K_x}{\omega^4 m_f}\left\{c_{43}^r + ic_{43}^i\right\}[\ddot{X}^g] \tag{4.95}$$

The transfer functions, $\tau_i(\omega)$, $\tau_c(\omega)$, $\tau_f(\omega)$ and $\tau_R(\omega)$, may be readily recognized from Equations (4.92) to (4.95) and written as follows:

$$\tau_i(\omega) = \frac{K_x}{\omega^4 m_f}\left\{c_{13}^r + ic_{13}^i\right\} \tag{4.96}$$

$$\tau_c(\omega) = \frac{K_x}{\omega^4 m_f}\left\{c_{23}^r + ic_{23}^i\right\} \tag{4.97}$$

$$\tau_f(\omega) = \frac{K_x}{\omega^4 m_f}\left\{c_{33}^r + ic_{33}^i\right\} \tag{4.98}$$

$$\tau_R(\omega) = \frac{K_x}{\omega^4 m_f}\left\{c_{43}^r + ic_{43}^i\right\} \tag{4.99}$$

Once the expressions for transfer functions correlating the wavelet coefficients of impulsive, convective and foundation displacements and rocking displacement (i.e. $\tau_i(\omega)$, $\tau_c(\omega)$, $\tau_f(\omega)$ and $\tau_R(\omega)$, respectively) to wavelet coefficients of ground accelerations are known, the transfer functions for hydrodynamic pressure, base shear and base moment (i.e. $\tau_H(\omega)$, $\tau_Q(\omega)$ and $\tau_M(\omega)$ respectively), as given by Equations (4.71), (4.82) and (4.85), may be computed. On substituting appropriate terms, dividing by $H\rho_l g$

(which is the maximum hydrostatic pressure at the tank base) using Equation (4.39) and subsequently simplifying, one may easily obtain the transfer function for hydrodynamic pressure coefficient, $\tau_{CH}(\omega)$, as follows:

$$\tau_{CH}(\omega) = -\left(\frac{\omega_{lf}}{\omega}\right)^2 \frac{1}{g}\left(\frac{1}{H/R}\right)[\{c_i(z)S_1 + c_c(R,z)S_2\} + i\{c_i(z)T_1 + c_c(R,z)T_2\}]$$
$$[\alpha_x + ia_0\beta_x] \tag{4.100}$$

In Equation (4.100), the expressions for S_1, S_2, T_1 and T_2 are as follows:

$$S_1 = c_{13}^r + c_{33}^r + h_i' c_{43}^r \tag{4.101}$$

$$S_2 = c_{23}^r + c_{33}^r + h_c' c_{43}^r \tag{4.102}$$

$$T_1 = c_{13}^i + c_{33}^i + h_i' c_{43}^i \tag{4.103}$$

$$T_2 = c_{23}^i + c_{33}^i + h_c' c_{43}^i \tag{4.104}$$

On substituting appropriate terms, dividing by $m_l g$ (which is the weight of the tank–liquid), using Equation (4.23) and simplifying further, one may get the transfer function for the tank base shear coefficient, $\tau_{CQ}(\omega)$, as follows:

$$\tau_{CQ}(\omega) = -\left(\frac{\omega_{lf}}{\omega}\right)^2 \frac{1}{g}\left[\left\{\frac{m_i}{m_l}S_1 + \frac{m_c}{m_l}S_2\right\} + i\left\{\frac{m_i}{m_l}T_1 + \frac{m_c}{m_l}T_2\right\}\right][\alpha_x + ia_0\beta_x] \tag{4.105}$$

In a similar way, the expression for the transfer function for the overturning base moment coefficient at the tank base may be obtained (in this case, we need to divide by $m_l Hg$ for making it dimensionless) as

$$\tau_{CM}(\omega) = -\left(\frac{\omega_{lf}}{\omega}\right)^2 \frac{1}{g}\left[\left\{\frac{m_i}{m_l}\frac{h_i}{H}S_1 + \frac{m_c}{m_l}\frac{h_i}{H}S_2\right\} + i\left\{\frac{m_i}{m_l}\frac{h_i}{H}T_1 + \frac{m_c}{m_l}\frac{h_c}{H}T_2\right\}\right]$$
$$[\alpha_x + ia_0\beta_x] \tag{4.106}$$

The expressions for hydrodynamic pressure coefficient, base shear coefficient and overturning base moment coefficient may be further simplified on collecting the real and imaginary components separately as follows:

$$\tau_{CH}(\omega) = -\left(\frac{\omega_{lf}}{\omega}\right)^2 \frac{1}{g}\left(\frac{1}{\frac{H}{R}}\right)[\{\alpha_x(c_i(z)S_1 + c_c(R,z)S_2) - a_0\beta_x(c_i(z)T_1$$

$$+ c_c(R,z)T_2)\} + i\{\alpha_x(c_i(z)T_1 + c_c(R,z)T_2) + a_0\beta_x(c_i(z)S_1 + c_c(R,z)S_2)\}] \quad (4.107)$$

$$\tau_{CQ}(\omega) = -\left(\frac{\omega_{lf}}{\omega}\right)^2 \frac{1}{g}\left[\left\{\alpha_x\left(\frac{m_i}{m_l}S_1 + \frac{m_c}{m_l}S_2\right) - a_0\beta_x\left(\frac{m_i}{m_l}T_1 + \frac{m_c}{m_l}T_2\right)\right\}\right.$$

$$\left. + i\left\{\alpha_x\left(\frac{m_i}{m_l}T_1 + \frac{m_c}{m_l}T_2\right) + a_0\beta_x\left(\frac{m_i}{m_l}S_1 + \frac{m_c}{m_l}S_2\right)\right\}\right] \quad (4.108)$$

$$\tau_{CM}(\omega) = -\left(\frac{\omega_{lf}}{\omega}\right)^2 \frac{1}{g}\left[\left\{\alpha_x\left(\frac{m_i}{m_l}\frac{h_i}{H}S_1 + \frac{m_c}{m_l}\frac{h_i}{H}S_2\right) - a_0\beta_x\left(\frac{m_i}{m_l}\frac{h_i}{H}T_1 + \frac{m_c}{m_l}\frac{h_c}{H}T_2\right)\right\}\right.$$

$$\left. + i\left\{\alpha_x\left(\frac{m_i}{m_l}\frac{h_i}{H}T_1 + \frac{m_c}{m_l}\frac{h_c}{H}T_2\right) + a_0\beta_x\left(\frac{m_i}{m_l}\frac{h_i}{H}S_1 + \frac{m_c}{m_l}\frac{h_i}{H}S_2\right)\right\}\right] \quad (4.109)$$

It may be noted here that the terms c^r_{mn} and c^i_{mn}, as used in the above equations, may be obtained from the matrix inversion process, which is not explained here in further detail. Thus, we now know the expressions of all transfer functions. Using these transfer functions, we may now start looking for the expected largest peak response of the tank–liquid–foundation system in the wavelet domain.

4.6 EXPECTED LARGEST PEAK RESPONSE

The expected largest peak response of the tank–liquid–foundation system is computed in this section with the help of statistical functionals of wavelet coefficients. First, we express the expected hydrodynamic pressure coefficients in terms of the expected ground motion process using the relevant transfer function. Subsequently, we evaluate the instantaneous energy in the mean square response of the pressure coefficients followed by evaluation of the power spectral density function. In this derivation, we follow the same procedure as shown in previous chapters.

On taking expectation of the square of the amplitude of the left- and right-hand sides of Equation (4.70), integrating over ω and using the orthogonality relation for the Littlewood–Paley wavelet basis function given by Equation (2.18), the relationship for a specific (jth) frequency band becomes

$$\sum_i E[\{W_\psi P_H(a_j,b_i)|^2] = \sum_i E[\{W_\psi \ddot{x}_g(a_j,b_i)|^2] \int_{-\infty}^{\infty} |\tau_H(\omega)|^2 \, |\psi_{a_j,b_i}(\omega)|^2 d\omega \tag{4.110}$$

The instantaneous energy for the mean square response in terms of hydrodynamic pressure coefficient (C_H) at a particular time instant ($t = b_i$) and corresponding to a specific band j (with dilation factor a_j) may be obtained using the time-localization property of wavelet coefficients and may be expressed as follows:

$$\int_{-\infty}^{\infty} E[|C_{iH}^{j}(\omega)|^2]d\omega = \frac{K\Delta b}{a_j} E\Big[\{W_\psi \ddot{x}_g(a_j,b_i)|^2\Big] \int_{-\infty}^{\infty} |\tau_{CH}(\omega)|^2 \, |\psi_{a_j,b_i}(\omega)|^2 \, d\omega \tag{4.111}$$

Equation (4.111) is an approximate relation. The instantaneous power spectral density function (S_{CH}) for the hydrodynamic pressure coefficient may now be obtained by summing up the average energy terms of the pressure coefficients (i.e. averaged over Δb) over all frequency bands, as shown below:

$$S_{CH}(\omega) = \sum_j \frac{E[|C_{iH}^{j}(\omega)|^2]}{\Delta b} = \sum_j \frac{K}{a_j} E[\{W_\psi \ddot{x}_g(a_j,b_i)|^2] \int_{-\infty}^{\infty} |\tau_{CH}(\omega)|^2 \, |\psi_{a_j,b_i}(\omega)|^2 \tag{4.112}$$

We know from previous discussions in Chapter 2 that the expressions for the zeroth, first and second moments of the PSDF are essential to evaluate the non-stationary peak factors of any responses. So, in this case we have to compute these values at every time point from their respective expressions given below:

$$m_0|_{t=b_i} = \sum_j \frac{K}{\pi(\sigma-1)} E[|W_\psi \ddot{x}_g(a_j,b_i)|^2] I_{0,j} \tag{4.113}$$

$$m_1|_{t=b_i} = \sum_j \frac{K}{\pi(\sigma-1)} E[|W_\psi \ddot{x}_g(a_j,b_i)|^2] I_{1,j} \tag{4.114}$$

$$m_2|_{t=b_i} = \sum_j \frac{K}{\pi(\sigma-1)} E[|W_\psi \ddot{x}_g(a_j,b_i)|^2] I_{2,j} \tag{4.115}$$

The expressions for the three moments of PSDF as shown in Equations (4.113) to (4.115) remain the same for hydrodynamic pressure coefficients, base shear coefficients and base moment coefficients. The difference lies in the expressions for $I_{0,j}$, $I_{1,j}$ and $I_{2,j}$. The expressions for $I_{0,j}$, $I_{1,j}$ and $I_{2,j}$ for hydrodynamic pressure coefficients are written below:

$$I_{0,j} = \int_{\frac{\pi}{a_j}}^{\frac{\sigma\pi}{a_j}} \left(\frac{\omega_{lf}}{\omega}\right)^4 \left(\frac{R}{gH}\right)^2 [\{\alpha_x(c_i(z)S_1 + c_c(R,z)S_2) - a_0\beta_x(c_i(z)T_1 + c_c(R,z)T_2)\}^2$$

$$+ \{\alpha_x(c_i(z)T_1 + c_c(R,z)T_2) + a_0\beta_x(c_i(z)S_1 + c_c(R,z)S_2)\}^2]d\omega \tag{4.116}$$

$$I_{1,j} = \int_{\frac{\pi}{a_j}}^{\frac{\sigma\pi}{a_j}} \left(\frac{\omega_{lf}}{\omega}\right)^4 \left(\frac{R}{gH}\right)^2 [\{\alpha_x(c_i(z)S_1 + c_c(R,z)S_2) - a_0\beta_x(c_i(z)T_1 + c_c(R,z)T_2)\}^2$$

$$+\{\alpha_x(c_i(z)T_1 + c_c(R,z)T_2) + a_0\beta_x(c_i(z)S_1 + c_c(R,z)S_2)\}^2]\omega d\omega \tag{4.117}$$

$$I_{2,j} = \int_{\frac{\pi}{a_j}}^{\frac{\sigma\pi}{a_j}} \left(\frac{\omega_{lf}}{\omega}\right)^4 \left(\frac{R}{gH}\right)^2 [\{\alpha_x(c_i(z)S_1 + c_c(R,z)S_2) - a_0\beta_x(c_i(z)T_1 + c_c(R,z)T_2)\}^2$$

$$+\{\alpha_x(c_i(z)T_1 + c_c(R,z)T_2) + a_0\beta_x(c_i(z)S_1 + c_c(R,z)S_2)\}^2]\,\omega^2 d\omega \tag{4.118}$$

In the same way, the expressions for the terms $I_{0,j}$, $I_{1,j}$ and $I_{2,j}$ for base shear coefficients may be written as

$$I_{0,j} = \int_{\frac{\pi}{a_j}}^{\frac{\sigma\pi}{a_j}} \left(\frac{\omega_{lf}}{\omega}\right)^4 \left(\frac{R}{gH}\right)^2 \left[\left\{\alpha_x\left(\frac{m_i}{m_l}S_1 + \frac{m_c}{m_l}S_2\right) - a_0\beta_x\left(\frac{m_i}{m_l}T_1 + \frac{m_c}{m_l}T_2\right)\right\}^2\right.$$

$$\left.+\left\{\alpha_x\left(\frac{m_i}{m_l}T_1 + \frac{m_c}{m_l}T_2\right) + a_0\beta_x\left(\frac{m_i}{m_l}S_1 + \frac{m_c}{m_l}S_2\right)\right\}^2\right]d\omega \tag{4.119}$$

$$I_{1,j} = \int_{\frac{\pi}{a_j}}^{\frac{\sigma\pi}{a_j}} \left(\frac{\omega_{lf}}{\omega}\right)^4 \left(\frac{R}{gH}\right)^2 \left[\left\{\alpha_x\left(\frac{m_i}{m_l}S_1 + \frac{m_c}{m_l}S_2\right) - a_0\beta_x\left(\frac{m_i}{m_l}T_1 + \frac{m_c}{m_l}T_2\right)\right\}^2 + \left\{\alpha_x\left(\frac{m_i}{m_l}T_1 + \frac{m_c}{m_l}T_2\right) + a_0\beta_x\left(\frac{m_i}{m_l}S_1 + \frac{m_c}{m_l}S_2\right)\right\}^2\right]\omega d\omega \tag{4.120}$$

$$I_{2,j} = \int_{\frac{\pi}{a_j}}^{\frac{\sigma\pi}{a_j}} \left(\frac{\omega_{lf}}{\omega}\right)^4 \left(\frac{R}{gH}\right)^2 \left[\left\{\alpha_x\left(\frac{m_i}{m_l}S_1 + \frac{m_c}{m_l}S_2\right) - a_0\beta_x\left(\frac{m_i}{m_l}T_1 + \frac{m_c}{m_l}T_2\right)\right\}^2 + \left\{\alpha_x\left(\frac{m_i}{m_l}T_1 + \frac{m_c}{m_l}T_2\right) + a_0\beta_x\left(\frac{m_i}{m_l}S_1 + \frac{m_c}{m_l}S_2\right)\right\}^2\right]\omega^2 d\omega \tag{4.121}$$

The expressions for the same terms in the case of base moment coefficients are given below:

$$I_{0,j} = \int_{\frac{\pi}{a_j}}^{\frac{\sigma\pi}{a_j}} \left(\frac{\omega_{lf}}{\omega}\right)^4 \left(\frac{R}{gH}\right)^2 \left[\left\{\alpha_x\left(\frac{m_i}{m_l}\frac{h_i}{H}S_1 + \frac{m_c}{m_l}\frac{h_i}{H}S_2\right) - a_0\beta_x\left(\frac{m_i}{m_l}\frac{h_i}{H}T_1 + \frac{m_c}{m_l}\frac{h_c}{H}T_2\right)\right\}^2 + \left\{\alpha_x\left(\frac{m_i}{m_l}\frac{h_i}{H}T_1 + \frac{m_c}{m_l}\frac{h_c}{H}T_2\right) + a_0\beta_x\left(\frac{m_i}{m_l}\frac{h_i}{H}S_1 + \frac{m_c}{m_l}\frac{h_i}{H}S_2\right)\right\}^2\right] d\omega \tag{4.122}$$

$$I_{1,j} = \int_{\frac{\pi}{a_j}}^{\frac{\sigma\pi}{a_j}} \left(\frac{\omega_{lf}}{\omega}\right)^4 \left(\frac{R}{gH}\right)^2 \left[\left\{\alpha_x\left(\frac{m_i}{m_l}\frac{h_i}{H}S_1 + \frac{m_c}{m_l}\frac{h_i}{H}S_2\right) - a_0\beta_x\left(\frac{m_i}{m_l}\frac{h_i}{H}T_1 + \frac{m_c}{m_l}\frac{h_c}{H}T_2\right)\right\}^2 + \left\{\alpha_x\left(\frac{m_i}{m_l}\frac{h_i}{H}T_1 + \frac{m_c}{m_l}\frac{h_c}{H}T_2\right) + a_0\beta_x\left(\frac{m_i}{m_l}\frac{h_i}{H}S_1 + \frac{m_c}{m_l}\frac{h_i}{H}S_2\right)\right\}^2\right]\omega d\omega \tag{4.123}$$

$$I_{2,j} = \int_{\pi/a_j}^{\sigma\pi/a_j} \left(\frac{\omega_{lf}}{\omega}\right)^4 \left(\frac{R}{gH}\right)^2 \left[\left[\left\{\alpha_x\left(\frac{m_i}{m_l}\frac{h_i}{H}S_1 + \frac{m_c}{m_l}\frac{h_i}{H}S_2\right) - a_0\beta_x\left(\frac{m_i}{m_l}\frac{h_i}{H}T_1 + \frac{m_c}{m_l}\frac{h_c}{H}T_2\right)\right\}^2\right.\right.$$

$$\left.+\left\{\alpha_x\left(\frac{m_i}{m_l}\frac{h_i}{H}T_1 + \frac{m_c}{m_l}\frac{h_c}{H}T_2\right) + a_0\beta_x\left(\frac{m_i}{m_l}\frac{h_i}{H}S_1 + \frac{m_c}{m_l}\frac{h_i}{H}S_2\right)\right\}^2\right]\omega^2 d\omega \qquad (4.124)$$

Now, it becomes easier to compute the nonstationary peak factors for the largest expected peak response. The probability that the process (nonstationary response in terms of hydrodynamic pressure or base shear or base moment) remains below the level x within a time interval $[0, T]$ is first computed from Equation (2.63). Subsequently, the time-dependent rate of the Poisson process is obtained using Equation (2.65), following which the expected value of the largest peak is obtained from Equation (2.67).

4.7 NUMERICAL EXAMPLE

The formulation of a multi-degree-of-freedom tank–fluid–foundation–soil system subjected to a random base excitation in the form of seismic ground motion is now clear to us. We will now try to apply the concept to solve the problem of a realistic tank model and obtain the nonstationary stochastic responses of the system. The objective is to carry out a parametric study on the stochastic responses of the tank–fluid–foundation system undergoing rocking motion to observe the effects of the variation in hydrodynamic pressure on the tank wall, shear developed at the tank–foundation interface and the overturning moment generated at the tank base due to the changes in liquid height, ratio of total liquid mass to foundation mass, slenderness of the tank and shear wave velocity in soil. This results in generation of some design charts that any user may develop following the formulation given in the previous sections. In the first set of examples, we will consider only impulsive liquid mass and ignore the convective mode of liquid vibration. However, soil interaction will be involved. The next set of examples will have soil–structure–fluid interaction analysis for the total structural model (considering both impulsive and first sloshing modes of fluid vibration). The results are all obtained from wavelet-based analysis of the models.

Table 4.1 Values of m_i/m_l in case of rigidly supported steel tanks for impulsive response

h/R	$H/R = 0.5$	$H/R = 1.0$	$H/R = 2.0$	$H/R = 3.0$
0.0005	0.300	0.549	0.694	0.695
0.001	0.304	0.554	0.700	0.703
0.002	0.312	0.565	0.711	0.713

Source: Veletsos, A. S., and Tang, Y., *Earthquake Eng. Struct. Dyn.*, 19, 473–496, 1990.

4.7.1 Impulsive response

For the analysis of the tank–fluid–soil interaction considering only the impulsive mode of liquid vibration, the tank with same linear elastic material properties as explained in Chapter 3 is considered. The values of material properties of linear elastic soil are as follows: Poisson's ratio = $\frac{1}{3}$ and mass density = 1500 kg/m^3. The properties of rigidly supported steel tanks for impulsive response are listed in Tables 4.1 to 4.3 [27]. These tables provide the values of m_i/m_l, h_i/H and h'_i/H, respectively, for various combinations of h/R and H/R ratios. These values are required to compute the responses of the structure as per the formulation laid out in previous

Table 4.2 Values of h_i/H in case of rigidly supported steel tanks for impulsive response

h/R	$H/R = 0.5$	$H/R = 1.0$	$H/R = 2.0$	$H/R = 3.0$
0.0005	0.385	0.415	0.491	0.547
0.001	0.387	0.417	0.492	0.548
0.002	0.392	0.420	0.494	0.549

Source: Veletsos, A. S., and Tang, Y., *Earthquake Eng. Struct. Dyn.*, 19, 473–496, 1990.

Table 4.3 Values of h'_i/H in case of rigidly supported steel tanks for impulsive response

h/R	$H/R = 0.5$	$H/R = 1.0$	$H/R = 2.0$	$H/R = 3.0$
0.0005	1.455	0.799	0.543	0.562
0.001	1.428	0.706	0.543	0.563
0.002	1.384	0.700	0.544	0.564

Source: Veletsos, A. S., and Tang, Y., *Earthquake Eng. Struct. Dyn.*, 19, 473–496, 1990.

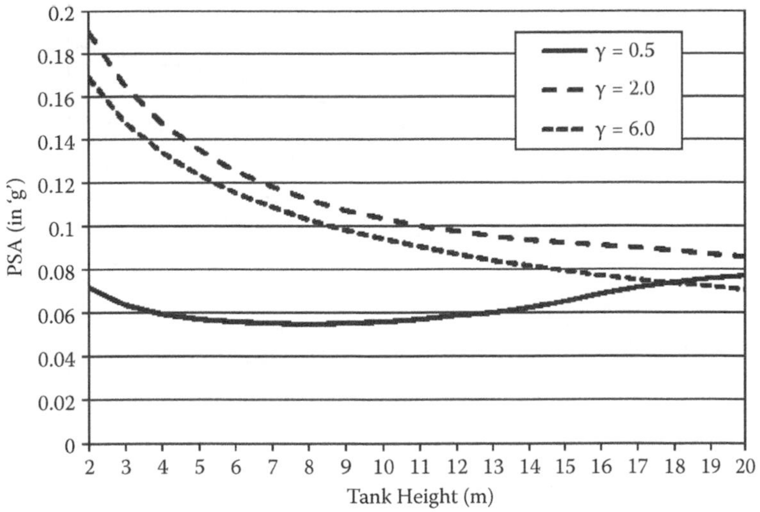

Figure 4.2 Variation of PSA with tank height for different mass ratios.

sections. In case of only the impulsive mode of liquid being considered, the convective mass of the liquid participating in vibrations is assumed to be zero, hence $m_c/m_l = 0$ and $m_i/m_l = 1.0$. The ratio of the density of water to steel is assumed to be 0.127. The damping for the impulsive part (ζ_i) is assumed as 5%. When only the impulsive part of the liquid mass and lateral motion of the system are considered (foundation thickness is assumed to be high to arrest rocking motion), the model reduces to a 2-DOF system (impulsive motion and foundation motion, both in the lateral direction only).

The impulsive modal mass heights, h_i and h'_i, are obtained from Tables 4.2 and 4.3 based on the choice of h/R (= 0.001 here) and H/R (= 0.5, 1 and 3 considered here) and the height of the tank assumed in the examples. The system is solved in the wavelet domain, and the pseudospectral acceleration (PSA) response spectrum is obtained for different parametric variations. Figure 4.2 shows the variation of the PSA response of the tank (with only impulsive liquid mass) with the tank height for different mass ratios, γ (mass of the impulsive liquid to the mass of the supporting foundation, i.e. m_i/m_l). Figures 4.3 and 4.4 demonstrate some design curves showing the variation of PSA responses with shear wave velocity (V_s) and lateral natural frequency of the foundation vibration (ω_{lf}), respectively, for different heights of the tank. Thus, more design curves could be generated in the same way, changing other parametric values.

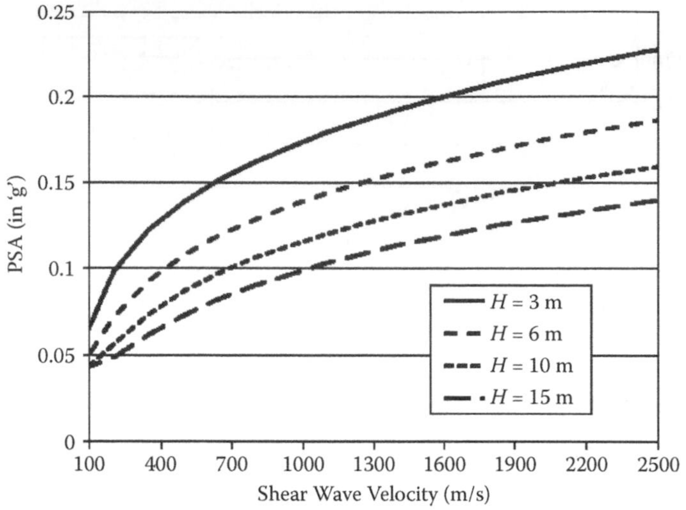

Figure 4.3 Variation of PSA with shear wave velocity for different tank heights.

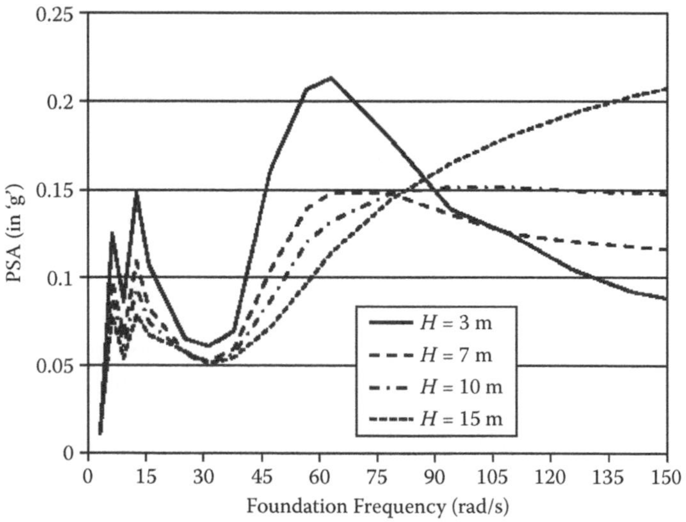

Figure 4.4 Variation of PSA with foundation frequency for different tank heights.

4.7.2 MDOF analysis results

Now, we make two important changes to our 2-DOF model. We introduce the first sloshing mode into the model and allow the rocking motion of the foundation simultaneously. The thickness of the foundation beneath the tank is assumed to be much less, so as to induce rocking motion into the system. The damping ratio corresponding to the first sloshing mode (i.e. ζ_c) is assumed to be 1%. The value of λ_1 is assumed to be 1.841 (vide Equation (4.16)). Hence, the system now turns into a MDOF (4-DOF) model. In this case, the corresponding values of m_i/m_l may be obtained from Table 4.1, and m_c/m_l is obtained from subtracting m_i/m_l from 1.0. The heights of lumped impulsive modal mass are obtained as usual from Tables 4.2 and 4.3. The heights of lumped convective modal mass may be found from Equations (4.8) to (4.10). Once these parameters are all known, the transfer functions and instantaneous moments and PSDFs as obtained from wavelet analysis may be computed and finally used to obtain the expected largest peak values of the response terms. The values of hydrodynamic pressure coefficients at different elevations of the tank wall thus found out are plotted in Figures 4.5 to 4.7, respectively, for variations in shear wave velocity

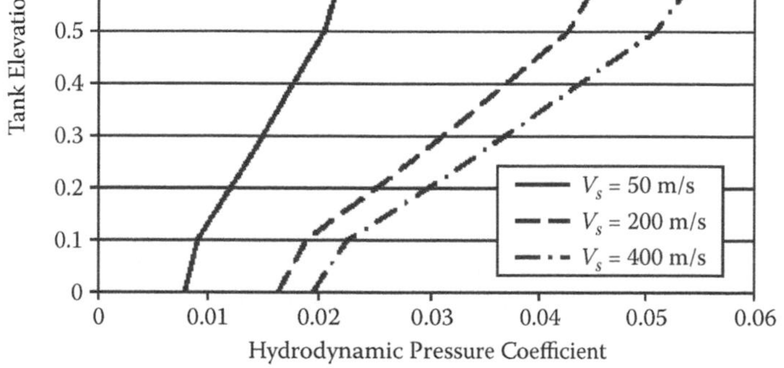

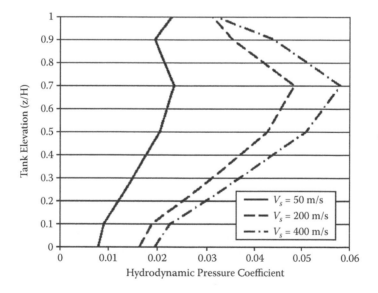

Figure 4.5 Pressure coefficients on tank wall for various shear wave velocities.

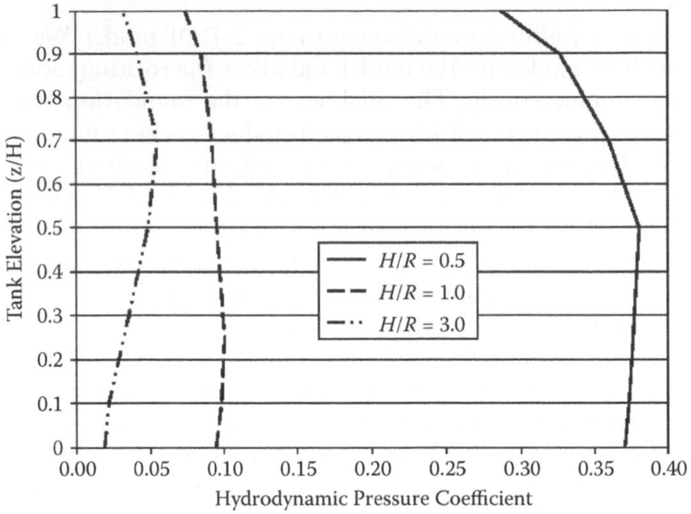

Figure 4.6 Pressure coeffcients on tank wall for various height–radius ratios of the tank.

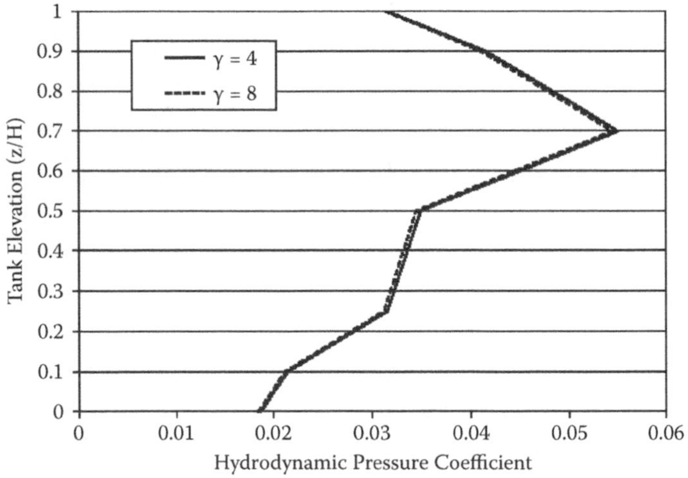

Figure 4.7 Pressure coefficients on tank wall for two mass ratios.

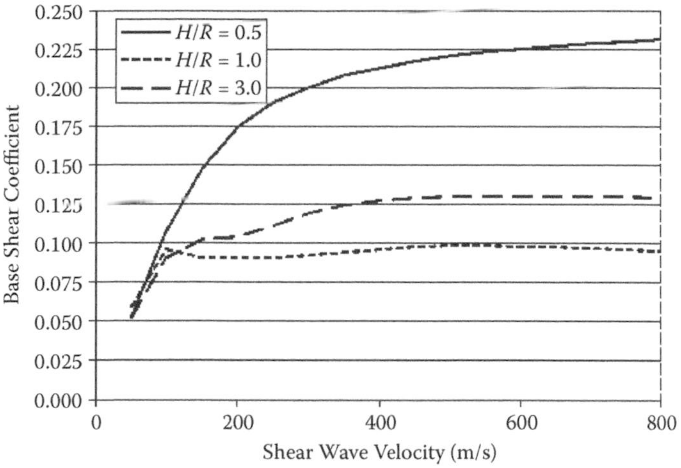

Figure 4.8 Shear coefficients at tank base for various shear wave velocities.

(50, 200 and 400 m/s), *H*/*R* ratios (0.5, 1 and 3) and ratio of total liquid mass to foundation mass (4 and 8). Figure 4.8 also depicts the variation in tank base shear coefficient with shear wave velocity in soil for different *H*/*R* ratios (0.5, 1 and 3). These figures may be used as design curves for different situations by structural designers, and similar others may be found out if necessary following the same wavelet-based approach, to include the ground motion nonstationarities.

Chapter 5

Wavelet-based nonstationary vibration analysis of a simple nonlinear system

The previous chapters have been devoted to the response analysis of linear systems subjected to nonstationary random vibrations. In these chapters, the basic idea of wavelets and the usefulness of wavelet-based analysis have been explained. The theory of wavelet-based analysis is used to develop and formulate the system equations of motions and transfer those to the wavelet domain and solve using wavelet coefficients. However, not all the structures behave in a way that could be well approximated by a linear system. In reality, the responses of a structure may deviate far from linear considerations [34–38]. The response of, say, buildings to earthquakes is a nonlinear dynamic problem. The readers should keep in mind that at this point we are talking about the nonlinearity present in the material model of the structure. However, it should also be borne in mind that we are not concerned about how to represent material nonlinearity; rather, we are focused only on how to formulate the equations of motion once we consider nonlinearity in our assumptions and how we solve the system dynamics using the wavelet analytical technique. The basic background theory regarding wavelets as discussed in previous chapters is used in this chapter too; however, the problem has become more complicated due to introduction of nonlinearity into the system. The main objective is to provide an insight into the formulation of the problem related to obtaining the nonlinear response of the system using the wavelet-based approach.

5.1 NONLINEAR SYSTEM

So far, we have considered linear structural systems in the analysis. However, most of the engineering structures include some degree of nonlinearity under real operating conditions. Though an assumption of a structure being linear is justified based on dynamic conditions and magnitude of external input excitation, nonlinear analysis of a structural system may become necessary as per design requirements that are aimed to improve the performance level, prediction accuracy, cost optimization, etc. Generally, different types of nonlinearities are introduced into the system due to stiffness, damping and friction effects, and these are usually dependent on amplitude of external force, velocity of motion and frequency of excitation. These are simplified and idealized so that the nonlinearity may be incorporated into subsequent simulation and analysis. The sources of nonlinearities may be broadly classified into material nonlinearity and geometric nonlinearity. To better understand the effect of nonlinear stiffness on the responses of structural systems, the Duffing oscillator is used to model the physical behaviour of the system with hardening (stiffening) or softening nonlinearity. The Duffing oscillator is the simplest appropriate nonlinear model available. A linear term and a nonlinear term are needed to describe the initial linear-elastic region of the stress-strain curve and the subsequent nonlinearities under larger deformations. Before going into the details of wavelet-based formulation for obtaining the response of a nonlinear structural system, let us first look into the basic theory of the Duffing oscillator, which will be subsequently used.

5.2 DUFFING OSCILLATOR

The single-degree-of-freedom (SDOF) system is modelled using the Duffing oscillator, and the response of the system subjected to nonstationary random excitations is investigated analytically. A damped Duffing oscillator with cubic nonlinearity in the restoring force is considered in the physical model of the system. The analytical treatment is carried out by the perturbation method because this is suitable for systems with weak nonlinearity, which is considered in the present analysis. Before going into the details of wavelet-based formulation and solution of the system response, a concise introduction to the Duffing oscillator and perturbation method may be useful to the readers.

The Duffing oscillator has become a classical paradigm for explaining the nonlinear behaviour of systems. Let us assume that a SDOF system is subjected to an external forcing function that is harmonic in nature. The relevant nondimensional Duffing equation is written as

$$\ddot{x}(t) + 2\zeta\dot{x}(t) + x(t) + \beta x^3(t) = F\cos(\Omega t) \tag{5.1}$$

In the above equation the terms $x(t)$, ζ and β on the left-hand side denote displacement (with respect to time), damping ratio and cubic stiffness parameter, respectively. The single dot and double dots represent velocity and acceleration, respectively. The terms F and Ω respectively represent the amplitude of the external force and the dimensionless frequency of the excitation (Ω is the ratio of the frequency of external force to the natural frequency of the system). If $\beta = 0$, Equation (5.1) reduces to the (forced) linear oscillator. For $\beta > 0$, the same equation represents a forced nonlinear oscillator, which is characterized by a nonlinear force-displacement curve. A positive value of β would correspond to a hardening (stiffening) spring, whereas a negative value would represent a softening spring. Now, one should note that Equation (5.1) stands for an externally excited Duffing oscillator where the excitation is external to the system. As this external forcing function is time dependent, the Duffing equation becomes a second-order nonautonomous system.

One may easily find that on assuming no external force (i.e. free vibration, so $F = 0$) and substituting $\beta = 0$ in Equation (5.1), the free vibration equation becomes

$$\ddot{x}(t) + 2\zeta\dot{x}(t) + x(t) = 0 \tag{5.2}$$

The initial conditions are assumed to be $x(0) = x_0$ and $\dot{x}(0) = \dot{x}_0$. The above equation may be solved assuming the solution to be of the following form:

$$x(t) = ke^{\lambda t} \tag{5.3}$$

On substituting this expression of $x(t)$ in Equation (5.2), using boundary conditions and making suitable simplifications (details may be followed from any textbook), the response for an underdamped system ($\zeta < 1$) may be obtained as

$$x(t) = Ke^{-\lambda t}\sin(\omega_d t + \varphi) \tag{5.4}$$

In the above equation, the expressions for the constant term K, the damped natural frequency ω_d and the phase angle φ are given as

$$K = \sqrt{x_0^2 + \left(\frac{\dot{x}_0 + \zeta x_0}{\omega_d}\right)^2} \tag{5.5}$$

$$\omega_d = \sqrt{1 - \zeta^2} \tag{5.6}$$

$$\varphi = \tan^{-1}\left(\frac{x_0\omega_d}{\dot{x}_0 + \zeta x_0}\right) \tag{5.7}$$

Equation (5.4) shows that the oscillations are decaying exponentially. When the external force is present without cubic nonlinearity in spring stiffness (so, the case becomes the same as linear damped forced vibration), the equation is written as

$$\ddot{x}(t) + 2\zeta\dot{x}(t) + x(t) = F\cos(\Omega t) \tag{5.8}$$

Using the same pair of initial conditions as used before for the case of free damped vibration, the nonhomogeneous equation (Equation (5.8)) gives a solution consisting of two parts, as written below:

$$x(t) = Ke^{-\lambda t}\cos(\omega_d t + \varphi) + T\cos(\Omega t + \phi) \tag{5.9}$$

The first part of the response, $x(t)$, is a homogenous part resembling an exponentially decaying response, and the second is the nonhomogeneous part resembling a steady-state response. The constant term ϕ denotes the phase shift with respect to the phase of the external force F and is given by the following relation:

$$\phi(\omega) = \begin{cases} -\tan^{-1}\left(\dfrac{2\zeta\Omega}{1-\Omega^2}\right), & \omega \le 1 \\ -\left(\pi - \tan^{-1}\left(\dfrac{2\zeta\Omega}{\Omega^2 - 1}\right)\right), & \omega > 1 \end{cases} \tag{5.10}$$

The term T is given as

$$T = \frac{|F|}{\sqrt{(1-\Omega^2)^2 + 4\zeta^2\Omega^2}} \tag{5.11}$$

The magnification factor, M, is obtained on dividing the amplitude of the nonhomogeneous part of the response given in Equation (5.9) (i.e. the term T) by the absolute amplitude of the excitation force, F:

$$M = \frac{1}{\sqrt{(1-\Omega^2)^2 + 4\zeta^2\Omega^2}} \tag{5.12}$$

There is an interesting observation for the magnification factor. For a specific damping ratio, the excitation frequency at which the maximum magnification takes place may be determined from the first- and second-order differentiations of M as follows:

$$\frac{dM(\Omega)}{d\Omega} = 0 \tag{5.13}$$

$$\frac{d^2M(\Omega)}{d\Omega^2} < 0 \tag{5.14}$$

Using Equations (5.12) to (5.14), it is found, after an easy exercise, that the maximum amplification (M) in the response is possible when $\Omega = \sqrt{1-2\zeta^2}$ for $0 < \zeta < \frac{1}{\sqrt{2}}$. This means that the lower the damping ratio, the closer the excitation frequency (causing maximum steady-state response) is to the natural frequency of the system. The change in the magnification factor due to the change in excitation frequency for various damping ratios is shown clearly in Figure 5.1.

It is also notable that when resonance occurs (i.e. $\Omega = 1$), the response of the system is unbounded, and hence the magnification factor also tends to become very high (infinity). A system with lower damping exhibits significant magnification over a shorter range of frequencies than a system with higher damping.

Now, let us turn our attention to the original Duffing equation, as given by Equation (5.1). A nonlinear term x^3 is seen in this equation, which means

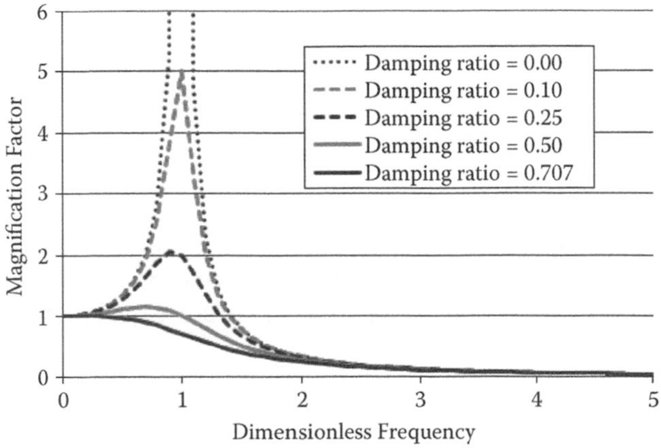

Figure 5.1 Variation in magnification factor due to change in excitation frequency for different values of damping ratios.

the linear superposition rule can no longer be used to obtain the forced response of the nonlinear system. The reader also has to keep in mind that the in the case of linear systems, the steady-state response does not depend on initial conditions. However, in the case of a nonlinear system the initial conditions influence the steady-state response. The maximum response also does not occur close to the natural frequency of the system if nonlinearity in the system is assumed. The cubic nonlinearity as considered can create resonance at a frequency that is away from the system natural frequency. A closed-form analytical solution to Equation (5.1) does not exist. However, some assumptions and approximations can be made to obtain an analytically approximate solution to the Duffing equation. In this case the method of perturbation will be used to solve the Duffing equation. The perturbation method and the solution of Duffing's equation are explained in the next sections.

5.3 PERTURBATION METHOD

The theory of perturbation is based on the concept of introducing small disturbances in the equation to solve it. It must be remembered that when the effect of a small parameter (simulating small disturbance) is small, the perturbation is termed regular perturbation; else, if the introduction of this small disturbance generates a large change in the solution, it is called a singular perturbation. Let us take a simple example on perturbation. We will find out the value of $\sqrt{50}$ using the perturbation method, i.e. by adding a small coefficient to 49. So, $\sqrt{50}$ can be rewritten as follows:

$$\begin{aligned}\sqrt{50} &= \sqrt{49+\varepsilon} = \sqrt{49+49*0.02} = \sqrt{49(1+0.02)}\\ &= 7*\sqrt{1.02} = 7*1.00995 = 7.06965\end{aligned}$$

Though the more approximate answer is 7.07107 (up to five places of decimal), the answer 7.06965 obtained from perturbation is quite acceptable.

Now, we look into a second example. We try to solve the following equation:

$$x-2=\varepsilon\cosh(x) \tag{5.15}$$

If $\varepsilon = 0$, $x = 2$. Assume $\varepsilon \ll 1$, and expand x as

$$x = x_0 + \varepsilon x_1 + \varepsilon^2 x_2 + \ldots \tag{5.16}$$

On substituting x in Equation (5.15) by its expansion as shown in Equation (5.2), the following equation is obtained:

$$(x_0 + \varepsilon x_1 + \varepsilon^2 x_2 + \ldots) - 2 = \varepsilon \cosh(x_0 + \varepsilon x_1 + \varepsilon^2 x_2 + \ldots) \quad (5.17)$$

The above equation can be solved separately for each order of ε. However, ε is assumed to be much smaller than 1, so the higher terms of ε can be ignored. Subsequently, solving only for the order of ε, one may obtain the following solution:

$$x \approx 2 + \varepsilon \cosh(2) \quad (5.18)$$

Thus, the basic principle of the regular perturbation of a general function $f(x)$ can be laid down in a few key points as follows:

1. Set $\varepsilon = 0$ and solve the resulting system (for definiteness).
2. The system now is perturbed by introducing nonzero ε, which is a small disturbance.
3. The solution to the new perturbed system can now be expanded as $f_0 + \varepsilon f_1 + \varepsilon^2 f_2 + \ldots$.
4. Subsequently, the governing equations are expanded in a series of ε. The terms with the same powers of ε are collected and solved individually until the desired solution is obtained.

5.4 SOLUTION OF THE DUFFING EQUATION

The Duffing oscillator was introduced originally to model the large-amplitude vibration modes of a steel beam when subjected to periodic forces. It is not possible to find an exact solution of the Duffing equation representing the dynamic behaviour of the Duffing oscillator. However, the solution can be approximated well by certain methods, of which the perturbation method is a widely used one. This method is described in the preceding section, and it is used to solve the equation pertaining to the Duffing oscillator as represented by Equation (5.1). Let us assume that the nonlinear system in consideration has weak nonlinearity, weak damping and weak forcing function. So, the damping ratio ζ and the nonlinear parameter β in Equation (5.1) may be reduced by a factor, say, $\varepsilon \ (\ll 1)$, and written as $\zeta' = \varepsilon\zeta$ and $\beta' = \varepsilon\beta$. The amplitude of the forcing function may similarly be written as $F'(= \varepsilon F)$. On using these terms, Equation (5.1) transforms to

$$\ddot{x}(t) + x(t) + \varepsilon\left\{2\zeta'\dot{x}(t) + \beta' x^3(t)\right\} = \varepsilon F' \cos(\Omega t) \quad (5.19)$$

The displacement response term, $x(t)$, is expanded as follows:

$$x(t) = x_0(t) + \varepsilon x_1(t) + \varepsilon^2 x_2(t) + \ldots \tag{5.20}$$

The expansion of $x(t)$ together with its differentiations can be used in Equation (5.19) to obtain an equation from which terms of the same powers of ε can be collected and the corresponding coefficients can be compared to obtain some more differential equations. These, in turn, can be solved using initial conditions of the considered system to obtain expressions for x_0, x_1, x_2, ... However, the maximum power of ε to be considered in the solution is based on the problem and accuracy of the solution sought after. Substituting the expressions of x_0, x_1, x_2, ... in Equation (5.20), the final solution of $x(t)$ is obtained.

5.5 NONLINEAR SYSTEM SUBJECTED TO RANDOM VIBRATION

A SDOF system is considered in the analysis and shown in Figure 5.2. The system is subjected to an input excitation that is nonstationary and random in nature. This SDOF model consists of an oscillator that is a combination of mass, spring and dashpot. The stiffness of the spring is assumed to behave nonlinearly. There are several options to consider nonlinearity in the spring. We consider here the cubic-type nonlinear spring force, $K[x(t) + \varepsilon x^3(t)]$, in which the term K denotes the linear static spring stiffness coefficient, ε is the nonlinear parameter and $x(t)$ denotes the displacement of the SDOF oscillator. This is similar to what is described as the Duffing oscillator in the foregoing section. Here, the first part in the expression, $Kx(t)$, is the linear spring element and the second part, $K\varepsilon x^3(t)$, is the cubic nonlinear spring element. If the value of the nonlinear parameter ε is less than zero, it resembles a softening spring. If the value of ε is positive, the spring element resembles a hardening (stiffening) spring. For the linear stiffness element, the force versus

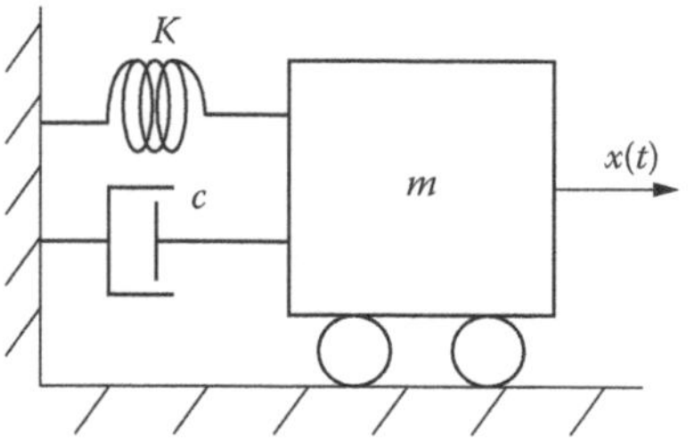

Figure 5.2 Nonlinear SDOF oscillator.

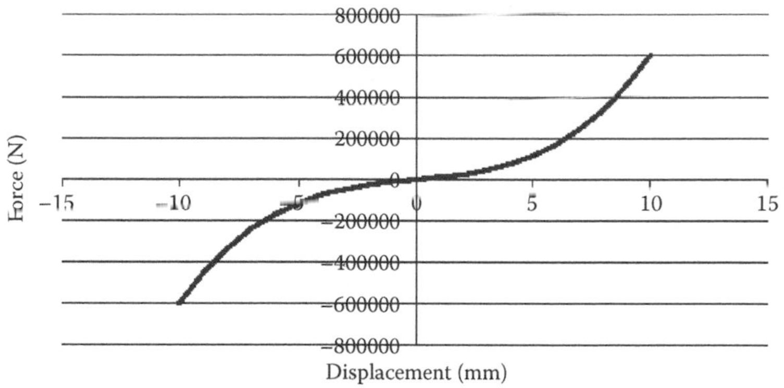

Figure 5.3 Hardening spring element with nonlinear (cubic) stiffness.

displacement curve becomes a straight line and its slope gives the stiffness constant K. However, with cubic nonlinearity, the force displacement relation no more remains a straight line. The stiffening (or softening) effect comes from geometry 'adaptation' to the applied loads because of redistribution of load, which occurs due to development of stresses as a reaction to applied loading. Figures 5.3 and 5.4 demonstrate the nonlinear (cubic) force-deformation relationship of hardening and softening spring elements, respectively. It is quite common to model the problems related to the vibration of beams, plates and shells by assuming cubic nonlinearity in the spring stiffness [39].

A system with skinny hysteresis can be well approximated by the combination of a linear viscous damper and a nonlinear cubic spring. The skinny hysteresis represents the fundamental nonlinear modal response of a structural system. Let us assume that the SDOF oscillator as shown in

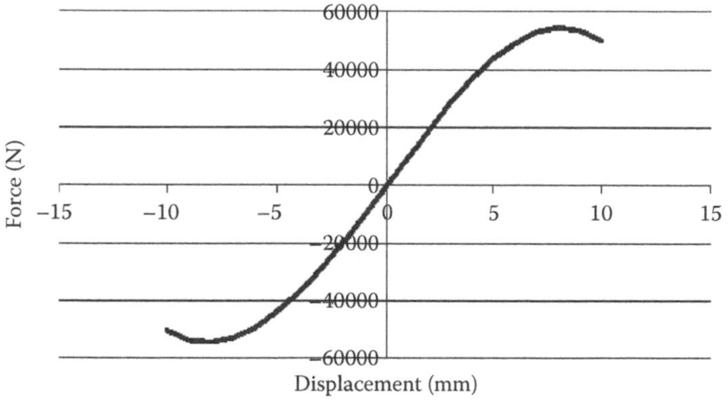

Figure 5.4 Softening spring element with nonlinear (cubic) stiffness.

Figure 5.2 represents a beam (with lumped mass m) and is subjected to random base acceleration $\ddot{x}_g(t)$. The damping is represented by introducing a viscous damping coefficient C. The equilibrium equation of the SDOF system is written as

$$m\ddot{x}(t)+K[x(t)+\varepsilon x^3(t)]+C\dot{x}(t)+m\ddot{x}_g(t)=0 \tag{5.21}$$

The overdots in Equation (5.21) represent differentiations with respect to time, t. On using the relations $K=m\omega_n^2$ and $C=2m\zeta\omega_n$ in Equation (5.21) and cancelling the common mass (m) terms from both sides, it is simplified to

$$\ddot{x}(t)+\omega_n^2x(t)+\varepsilon\omega_n^2x^3(t)+2\zeta\omega_n\dot{x}(t)+\ddot{x}_g(t)=0 \tag{5.22}$$

The terms ζ and ω_n, respectively, denote the damping ratio of the linear component of the SDOF oscillator (viscous damper) and the natural frequency of the SDOF system. The following pair of substitutions is to be made in order to transform the above equation into nondimensional form:

$$T=\omega_n t \tag{5.23}$$

and

$$\Delta(T)=\frac{\omega_n^2x(t)}{\ddot{x}_{gm}} \tag{5.24}$$

In the above equation, $\ddot{x}_{gm}$ denotes the absolute value of the maximum ground acceleration (also otherwise called peak ground acceleration (PGA)). From Equations (5.23) and (5.24) the following expressions for $x(t)$, $\dot{x}(t)$ and $\ddot{x}(t)$ are derived:

$$x(t)=\Delta(T).\frac{\ddot{x}_{gm}}{\omega_n^2} \tag{5.25}$$

$$\dot{x}(t)=\frac{d}{dt}(x(t))=\left\{\frac{d}{dT}\left(\Delta(T).\frac{\ddot{x}_{gm}}{\omega_n^2}\right)\right\}\frac{dT}{dt}=\left\{\frac{\ddot{x}_{gm}}{\omega_n^2}.\frac{d\Delta(T)}{dT}\right\}.\omega_n \tag{5.26}$$

$$\ddot{x}(t)=\frac{d}{dt}(\dot{x}(t))=\frac{d}{dT}\left[\left\{\frac{\ddot{x}_{gm}}{\omega_n^2}.\frac{d\Delta(T)}{dT}\right\}.\omega_n\right]\frac{dT}{dt}=\left\{\frac{\ddot{x}_{gm}}{\omega_n^2}.\frac{d^2\Delta(T)}{dT^2}\right\}.\omega_n^2 \tag{5.27}$$

On substituting Equations (5.25) to (5.27) in Equation (5.22), the system equilibrium equation reduces to the following nondimensional form:

$$\frac{d^2\Delta(T)}{dT^2}+\Delta(T)+\kappa\ \Delta^3(T)+2\zeta\frac{d\Delta(T)}{dT}+\frac{\ddot{x}_g(T)}{\ddot{x}_{gm}}=0 \tag{5.28}$$

In fact, the term T in the above equations represents a nondimensional time parameter, and the term Δ stands for the nondimensional displacement parameter. The term κ in Equation (5.28) is the nondimensional nonlinear parameter evolved in the course of simplification of equations and has the following expression:

$$\kappa=\varepsilon\left(\frac{\ddot{x}_{gm}}{\omega_n}\right)^2 \tag{5.29}$$

In order to solve Equation (5.28), the perturbation method is used here. Let us perturb the nondimensional displacement term, Δ, of the oscillator as shown below:

$$\Delta(T)=\Delta_0(T)+\kappa\Delta_1(T)+\kappa^2\Delta_2(T)+\kappa^3\Delta_3(T)+\ldots \tag{5.30}$$

As the model in consideration is subjected to nonstationary stochastic excitation at the base, the concept of the wavelet analytical technique will now be used to solve the dynamic equation. On taking the wavelet transform of the terms on both sides of Equation (5.30), one may get

$$\begin{aligned}W_\psi\Delta(a_j\omega_n,b_i\omega_n)&=W_\psi\Delta_0(a_j\omega_n,b_i\omega_n)+\kappa W_\psi\Delta_1(a_j\omega_n,b_i\omega_n)\\&\quad+\kappa^2W_\psi\Delta_2(a_j\omega_n,b_i\omega_n)+\kappa^3W_\psi\Delta_3(a_j\omega_n,b_i\omega_n)+\ldots\end{aligned} \tag{5.31}$$

As we have seen in previous chapters, it is necessary to compute the energy given by the term $E[\{W_\psi\Delta(a_j\omega_n,b_i\omega_n)\}^2]$ to obtain the nonlinear stochastic response of the given SDOF system. As the value of κ is small due to small disturbance (perturbation) introduced into the system, values greater than 1 in the order of κ will be ignored. Hence, on considering only the first two terms on the right-hand side of the Equation (5.31) and subsequently taking squares of all terms on both sides and then their expectations, one may get

$$E\left[\{W_\psi\Delta(a_j',b_i')\}^2\right]=E\left[\{W_\psi\Delta_0(a_j',b_i')\}^2\right]+2\kappa E\left[\{W_\psi\Delta_0(a_j',b_i')\}\{W_\psi\Delta_1(a_j',b_i')\}\right] \tag{5.32}$$

In Equation (5.32), the terms a_j' and b_j' have the following expressions:

$$a_j' = a_j \omega_n \tag{5.33}$$

$$b_i' = b_i \omega_n \tag{5.34}$$

It is important here to note that Equation (5.28) has a cubic term of Δ (i.e. $\Delta^3(T)$), which should also be replaced with wavelet coefficients. On cubing all terms in Equation (5.30), considering only the first-order term of κ and using substitutions as given in Equations (5.33) and (5.34), one may get

$$\Delta^3(T) = \Delta_0^3(T) + 3\kappa\Delta_0^2(T)\Delta_1(T) \tag{5.35}$$

On taking the wavelet transform of both sides of Equation (5.35), the following equation is obtained:

$$W_\psi \Delta^3(a_j', b_i') = W_\psi \Delta_0^3(a_j', b_i') + 3\kappa W_\psi \left[\Delta_0^2(a_j', b_i').\Delta_1(a_j', b_i')\right] \tag{5.36}$$

We have seen before that any response may be expressed in terms of wavelet coefficients. In this case, the term $\Delta^3(T)$ may be expanded in terms of wavelet coefficients as

$$\Delta^3(T) = \sum_i \sum_j \frac{K\Delta b^*}{a_j'} W_\psi \Delta^3(a_j', b_i') \psi_{a_j', b_i'}(T) \tag{5.37}$$

The term Δb^* in Equation (5.37) is given as

$$\Delta b^* = \Delta b.\omega_n \tag{5.38}$$

The term $W_\psi \Delta^3(a_j', b_i')$ on the right-hand side of Equation (5.37) can be replaced by its expression from Equation (5.36), and this leads to obtaining the expression of the nondimensional displacement term, Δ, dependent on nonlinear parameter κ as follows:

$$\begin{aligned}\Delta^3(T) = {} & \sum_i \sum_j \frac{K\Delta b^*}{a_j'} W_\psi \Delta_0^3(a_j', b_i') \psi_{a_j', b_i'}(T) \\ & + 3\kappa \sum_i \sum_j \frac{K\Delta b^*}{a_j'} W_\psi [\Delta_0^2(a_j', b_i').\Delta_1(a_j', b_i')] \psi_{a_j', b_i'}(T)\end{aligned} \tag{5.39}$$

Equation (5.39) is necessary for solving Equation (5.28). For the same reason, we now need to obtain the expansion of $\Delta(T)$ in terms of wavelet coefficients. This is written below:

$$\begin{aligned}\Delta(T) = &\sum_i\sum_j \frac{K\Delta b^*}{a'_j} W_\psi \Delta_0(a'_j, b'_i)\psi_{a'_j,b'_i}(T) \\ &+\kappa\sum_i\sum_j \frac{K\Delta b^*}{a'_j} W_\psi \Delta_1(a'_j, b'_i)\psi_{a'_j,b'_i}(T)\end{aligned} \tag{5.40}$$

The expression of seismic ground acceleration can be expressed in terms of wavelet coefficients and is shown in Chapter 3. However, for a quick recapitulation, let us write the expression once more;

$$\ddot{x}_g(t) = \sum_i\sum_j \frac{K\Delta b}{a_j} W_\psi\, \ddot{x}_g(a_j, b_i)\psi_{a_j,b_i}(t) \tag{5.41}$$

Using nondimensional terms, the ground acceleration may otherwise be written as

$$\ddot{x}''_g(T) = \sum_i\sum_j \frac{K\Delta b^*}{a'_j} W_\psi\, \ddot{x}_g(a'_j, b'_i)\psi_{a_j,b_i}(T) \tag{5.42}$$

On substituting Equations (5.39), (5.40) and (5.42) in Equation (5.28) and neglecting higher-order terms of κ (i.e. κ^2 and higher), the following wavelet domain equation is obtained:

$$\begin{aligned}&\sum_i\sum_j \frac{K\Delta b^*}{a'_j} W_\psi \Delta_0(a'_j, b'_i)\ddot{\psi}_{a'_j,b'_i}(T) + \kappa\sum_i\sum_j \frac{K\Delta b^*}{a'_j} W_\psi \Delta_1(a'_j, b'_i)\ddot{\psi}_{a'_j,b'_i}(T) \\ &+\sum_i\sum_j \frac{K\Delta b^*}{a'_j} W_\psi \Delta_0(a'_j, b'_i)\psi_{a'_j,b'_i}(T) + \kappa\sum_i\sum_j \frac{K\Delta b^*}{a'_j} W_\psi \Delta_1(a'_j, b'_i)\psi_{a'_j,b'_i}(T) \\ &+\kappa\sum_i\sum_j \frac{K\Delta b^*}{a'_j} W_\psi \Delta_0^3(a'_j, b'_i)\psi_{a'_j,b'_i}(T) + 2\zeta\sum_i\sum_j \frac{K\Delta b^*}{a'_j} W_\psi \Delta_0(a'_j, b'_i)\dot{\psi}_{a'_j,b'_i}(T) \\ &+2\zeta\kappa\sum_i\sum_j \frac{K\Delta b^*}{a'_j} W_\psi \Delta_1(a'_j, b'_i)\dot{\psi}_{a'_j,b'_i}(T) \\ &+\sum_i\sum_j \frac{K\Delta b^*}{a'_j} W_\psi \ddot{x}''_{gm}(a'_j, b'_i)\psi_{a'_j,b'_i}(T) = 0\end{aligned} \tag{5.43}$$

In Equation (5.43), the term $\ddot{x}''_{gm}$ is given as

$$\ddot{x}''_{gm}(T) = \frac{\ddot{x}''_g(T)}{\ddot{x}_{gm}} \tag{5.44}$$

There are two unknowns in Equation (5.43): Δ_0 and Δ_1. The method of perturbation is used for the solution; hence, these two unknown terms can be found from two separate equations that can be formed by collecting the terms associated with nonlinear perturbation parameters κ^0 and κ^1 from both sides of Equation (5.43) and equating them. Thus, on equating the coefficients of κ^0 from Equation (5.43), one may get

$$\sum_i\sum_j \frac{K\Delta b^*}{a'_j} W_\psi \Delta_0(a'_j,b'_i)\ddot{\psi}_{a'_j,b'_i}(T) + \sum_i\sum_j \frac{K\Delta b^*}{a'_j} W_\psi \Delta_0(a'_j,b'_i)\psi_{a'_j,b'_i}(T)$$
$$+2\zeta\sum_i\sum_j \frac{K\Delta b^*}{a'_j} W_\psi \Delta_0(a'_j,b'_i)\dot{\psi}_{a'_j,b'_i}(T) = -\sum_i\sum_j \frac{K\Delta b^*}{a'_j} W_\psi\, \ddot{x}''_{gm}(a'_j,b'_i)\psi_{a'_j,b'_i}(T) \tag{5.45}$$

Similarly, equating the coefficients of κ^1, one can easily find the other equation as

$$\sum_i\sum_j \frac{K\Delta b^*}{a'_j} W_\psi \Delta_1(a'_j,b'_i)\ddot{\psi}_{a'_j,b'_i}(T) + \sum_i\sum_j \frac{K\Delta b^*}{a'_j} W_\psi \Delta_1(a'_j,b'_i)\psi_{a'_j,b'_i}(T)$$
$$+\sum_i\sum_j \frac{K\Delta b^*}{a'_j} W_\psi \Delta_0^3(a'_j,b'_i)\psi_{a'_j,b'_i}(T) + 2\zeta\sum_i\sum_j \frac{K\Delta b^*}{a'_j} W_\psi \Delta_1(a'_j,b'_i)\dot{\psi}_{a'_j,b'_i}(T) = 0 \tag{5.46}$$

Let us deal with the above two wavelet domain equations one by one. On Fourier transforming Equation (5.45), the following equation is obtained:

$$\sum_i\sum_j \frac{K\Delta b^*}{a'_j} W_\psi \Delta_0(a'_j,b'_i)\hat{\psi}_{a'_j,b'_i}(\omega)(-\omega^2) + \sum_i\sum_j \frac{K\Delta b^*}{a'_j} W_\psi \Delta_0(a'_j,b'_i)\hat{\psi}_{a'_j,b'_i}(\omega)$$
$$+2\zeta\sum_i\sum_j \frac{K\Delta b^*}{a'_j} W_\psi \Delta_0(a'_j,b'_i)\hat{\psi}_{a'_j,b'_i}(\omega)(i\omega) = -\sum_i\sum_j \frac{K\Delta b^*}{a'_j} W_\psi\, \ddot{x}''_{gm}(a'_j,b'_i)\hat{\psi}_{a'_j,b'_i}(\omega) \tag{5.47}$$

On collecting similar terms, Equation (5.47) takes the following form:

$$\begin{aligned}&\sum_i\sum_j \frac{K\Delta b^*}{a'_j} W_\psi \Delta_0(a'_j,b'_i)\hat{\psi}_{a'_j,b'_i}(\omega)[1-\omega^2+i2\zeta\omega]\\ &\quad -\sum_i\sum_j \frac{K\Delta b^*}{a'_j} W_\psi \ddot{x}''_{gm}(a'_j,b'_i)\hat{\psi}_{a'_j,b'_i}(\omega)\end{aligned} \tag{5.48}$$

Now, first multiply all terms in Equation (5.48) by $\hat{\psi}^*_{a_k,b_l}(\omega)$ and subsequently use the relation defined in Equation (2.40) in Chapter 2 to arrive at the following equation:

$$[1-\omega^2+i2\zeta\omega]\sum_i W_\psi \Delta_0(a'_j,b'_i)\hat{\psi}_{a'_j,b'_i}(\omega) = -\sum_i W_\psi \ddot{x}''_{gm}(a'_j,b'_i)\hat{\psi}_{a'_j,b'_i}(\omega) \tag{5.49}$$

from which the transfer function $H_1(\omega)$ may be derived as shown below:

$$\sum_i W_\psi \Delta_0(a'_j,b'_i)\hat{\psi}_{a'_j,b'_i}(\omega) = H_1(\omega)\sum_i W_\psi \ddot{x}''_{gm}(a'_j,b'_i)\hat{\psi}_{a'_j,b'_i}(\omega) \tag{5.50}$$

where

$$H_1(\omega) = \frac{-1}{[(1-\omega^2)+i(2\zeta\omega)]} \tag{5.51}$$

This transfer function defines the relation between the nondimensional displacement term Δ_0 and the ground acceleration in terms of wavelet coefficients. On proceeding in a similar manner with the other wavelet domain equation (Equation (5.46)), the following transfer function ($H_2(\omega)$) relating the wavelet coefficients of Δ_1 to the wavelet coefficients of Δ_0^3 can be obtained:

$$\sum_i W_\psi \Delta_1(a'_j,b'_i)\hat{\psi}_{a'_j,b'_i}(\omega) = H_2(\omega)\sum_i W_\psi \Delta_0^3(a'_j,b'_i)\hat{\psi}_{a'_j,b'_i}(\omega) \tag{5.52}$$

where

$$H_2(\omega) = \frac{-1}{[(1-\omega^2)+i(2\zeta\omega)]} \tag{5.53}$$

It may be noted that these two transfer functions are the same, i.e. $H_1(\omega) = H_2(\omega)$. It is seen from Equation (5.52) that term Δ_1 is related to Δ_0^3, but the expectation is that Δ_1 should be related to the ground acceleration term $\ddot{x}''_{gm}$ in a similar way as the term Δ_0 is related to $\ddot{x}''_{gm}$ (as seen in Equation (5.50)). In order to accomplish this, we need to put up some more effort, which is elucidated in the next few steps.

In the first step, Equation (5.52) is multiplied by $\Sigma_k W_\psi \Delta_0(a'_j, b'_k)\hat{\psi}^*_{a'_j,b'_k}(\omega)$ to get

$$\sum_i \left[\left\{ W_\psi \Delta_1(a'_j, b'_i)\left(\sqrt{a'_j}\,\hat{\psi}(a'_j\omega)e^{-ib'_i\omega}\right)\right\} . \left\{ \sum_k W_\psi \Delta_0(a'_j, b'_k)\left(\sqrt{a'_j}\,\hat{\psi}^*(a'_j\omega)e^{ib'_k\omega}\right)\right\}\right]$$
$$= H_2(\omega)\sum_i \left[\left\{ W_\psi \Delta_0^3(a'_j, b'_i)\left(\sqrt{a'_j}\,\hat{\psi}(a'_j\omega)e^{-ib'_i\omega}\right)\right\}\right.$$
$$\left.\times \left\{ \sum_k W_\psi \Delta_0(a'_j, b'_k)\left(\sqrt{a'_j}\,\hat{\psi}^*(a'_j\omega)e^{ib'_k\omega}\right)\right\}\right] \tag{5.54}$$

The above equation can be expanded as

$$\sum_i \left[W_\psi \Delta_1(a'_j, b'_i) \cdot W_\psi \Delta_0(a'_j, b'_i) a'_j \,|\hat{\psi}(a'_j\omega)|^2 \right]$$
$$+ \sum_i \left[W_\psi \Delta_1(a'_j, b'_i) \sum_{k \neq i} W_\psi \Delta_0(a'_j, b'_k) a'_j \,|\hat{\psi}(a'_j\omega)|^2 \; e - (b'_i - b'_k)\omega \right]$$
$$= \sum_i \left[H_2(\omega) W_\psi \Delta_0^3(a'_j, b'_i) W_\psi \Delta_0(a'_j, b'_i) a'_j \,|\hat{\psi}(a'_j\omega)|^2 \right]$$
$$+ \left[H_2(\omega) W_\psi \Delta_0^3(a'_j, b'_i) . \sum_{k \neq i} W_\psi \Delta_0(a'_j, b'_k) a'_j \,|\,\hat{\psi}(a'_j\omega)\,|^2 \; e^{-(b'_i - b'_k)\omega} \right] \tag{5.55}$$

There are two operations that should now be performed simultaneously to simplify Equation (5.55). The first operation is to use the following relation to replace the term:

$$|\hat{\psi}(a'_j\omega)|^2 = \frac{|\hat{\psi}_{a'_j,b'_i}(\omega)|^2}{a'_j} \tag{5.56}$$

The remaining operation is to integrate both sides of Equation (5.55) over ω so that the double summation terms (the second term on both sides of the equation) disappear (see Section 2.5 for details). The resulting equation looks like the following:

$$\sum_{i(=k)}\left[W_\psi\Delta_1(a_j',b_i').W_\psi\Delta_0(a_j',b_i')a_j'\left|\hat{\psi}(a_j'\omega)\right|^2\right]$$
$$=\sum_{i(=k)}\left[H_2(\omega)W_\psi\Delta_0^3(a_j',b_i').W_\psi\Delta_0(a_j',b_i')a_j'\left|\psi(a_j'\omega)\right|^?\right] \tag{5.57}$$

At this point, we shall use the following relation:

$$\int\int E[W_\psi f.W_\psi g]da.db=\int E[g.f]dt \tag{5.58}$$

Using the discretized version of the above relation, we can express the product of wavelet coefficients of the linear and cubic terms of Δ_0 (assume $f=\Delta_0$ and $g=\Delta_0^3$), as seen in the right-hand side of Equation (5.57), as shown below:

$$\sum_i\sum_j\frac{K\Delta b^*}{a_j'}E[W_\psi\Delta_0^3(a_j',b_i').W_\psi\Delta_0(a_j',b_i')]=\int E[\Delta_0^4]dt \tag{5.59}$$

Now, considering the instantaneous relationship at every time point b_i', it can be written

$$E[\Delta_0^4]_{|t=b_i'}=\sum_j\frac{K}{a_j'}E[W_\psi\Delta_0^3(a_j',b_i').W_\psi\Delta_0(a_j',b_i')] \tag{5.60}$$

The assumption here is that the nondimensional displacement term, Δ_0, is a locally Gaussian nonstationary random process for which we have the following equation:

$$E\left[\Delta_0^4\right]_{|t=b_i'}=3\left[E\left\{\Delta_0^2\right\}_{|t=b_i'}\right]^2 \tag{5.61}$$

Let us suppose that the terms f and g are both equal to Δ_0^2. In that case, Equation (5.58) gives rise to the following discretized version:

$$\sum_i\sum_j\frac{K\Delta b^*}{a_j'}E\left[\left\{W_\psi\Delta_0^2(a_j',b_i')\right\}^2\right]=\int E\left[\left(\Delta_0^2\right)^2\right]dt \tag{5.62}$$

Using Equations (5.59), (5.61) and (5.62), one may write

$$\sum_j \frac{K\Delta b^*}{a'_j} E\left[W_\psi \Delta_0^3(a'_j, b'_i).W_\psi \Delta_0(a'_j, b'_i)\right] = 3\Delta b^* \left[\sum_j \frac{K}{a'_j} E\left[\left\{W_\psi \Delta_0^2(a'_j, b'_i)\right\}^2\right]\right]^2 \tag{5.63}$$

Here, the right-hand side of Equation (5.63) involving the term Δ_0^2 needs to be replaced with the term containing the ground motion term, $\ddot{x}''_{gm}(a'_j, b'_i)$. To do so, Equation (5.50) is used and the resulting expression for $\{W_\psi \Delta_0^2(a'_j, b'_i)\}^2$ becomes

$$\left\{W_\psi \Delta_0^2(a'_j, b'_i)\right\}^2 = \left\{\left(\left|H_1(\omega)\right|\right)^2 \left(W_\psi \ddot{x}''_{gm}(a'_j, b'_i)\right)^2 \left(\left|\hat{\psi}_{a'_j, b'_i}(\omega)\right|\right)^2\right\}^2 \tag{5.64}$$

The term $(|H_1(\omega)|)^2$ has the following expression:

$$\left|H_1(\omega)\right|^2 = \frac{1}{\left(1-\omega^2\right)^2 + (2\zeta\omega)^2} \tag{5.65}$$

Using Equation (5.64) in Equation (5.63) and integrating over ω, one may write

$$\sum_j \int_{-\infty}^{\infty} \frac{K\Delta b^*}{a'_j} E\left[W_\psi \Delta_0^3(a'_j, b'_i).W_\psi \Delta_0(a'_j, b'_i)\right]\left(\left|\hat{\psi}_{a'_j, b'_i}(\omega)\right|\right)^2 d\omega$$

$$= 3\Delta b^* \left[\sum_j \int_{-\infty}^{\infty} \frac{K}{a'_j} E\left[\left\{W_\psi \ddot{x}''_{gm}(a'_j, b'_i)\right\}^2\right].\left|H_1(\omega)\right|^2.\left|\hat{\psi}_{a'_j, b'_i}(\omega)\right|^2 d\omega\right]^2 \tag{5.66}$$

Using the following relationship (to break down the whole square term on the right-hand side of Equation (5.66) into the product of two separate terms),

$$\Phi = \sum_k \int_{-\infty}^{\infty} \frac{K}{a_k} E\left[\left\{W_\psi \ddot{x}''_{gm}(a'_k, b'_i)\right\}^2\right].\left|H_1(\omega)\right|^2.\left|\hat{\psi}_{a'_k, b'_i}(\omega)\right|^2 d\omega \tag{5.67}$$

Equation (5.66) may also be written in the form

$$\sum_{j}\int_{-\infty}^{\infty}\frac{K\Delta b^{*}}{a_j'}E\left[W_{\psi}\Delta_0^3(a_j',b_i').W_{\psi}\Delta_0(a_j',b_i')\right]\left(\left|\hat{\psi}_{a_j',b_i'}(\omega)\right|\right)^2 d\omega$$
$$=3\sum_{j}\int_{-\infty}^{\infty}\frac{K\Delta b^{*}}{a_j'}E\left[\left\{W_{\psi}\ddot{x}_{gm}''(a_j',b_i')\right\}^2\right].\left|H_1(\omega)\right|^2.\left|\psi_{a_j',b_i'}(\omega)\right|^2.\Phi\, d\omega \tag{5.68}$$

For the readers, there are some important points to note here. First, the value of σ is close to 1 in the case of nonstationary ground motions (viz. Section 2.4), and the integrands in Equation (5.68) also get closer to 1. Hence, the integral sign may be dropped to obtain a point-wise equality in the relationship. In addition, as per the basic assumption that the frequency bands are nonoverlapping in nature, the expressions under the summation sign on both sides of Equation (5.68) are equal. Thus, on multiplying all terms of this equation by the term $H_2(\omega)$ and summing up over all time instants, the relation between the nondimensional displacement term and the ground motion for a specific frequency band may be expressed as follows:

$$\sum_{i}E\left[W_{\psi}\Delta_0^3(a_j',b_i').W_{\psi}\Delta_0(a_j',b_i')\right]H_2(\omega)\left|\hat{\psi}_{a_j',b_i'}(\omega)\right|^2$$
$$=3\sum_{i}\Phi E\left[\left\{W_{\psi}\ddot{x}_{gm}''(a_j',b_i')\right\}^2\right].\left|H_1(\omega)\right|^2.H_2(\omega).\left|\hat{\psi}_{a_j',b_i'}(\omega)\right|^2 \tag{5.69}$$

Combining Equations (5.57) and (5.69), the cubic nondimensional displacement term may be avoided as follows:

$$\sum_{i}E\left[W_{\psi}\Delta_1(a_j',b_i').W_{\psi}\Delta_0(a_j',b_i')\right]\left|\hat{\psi}_{a_j',b_i'}(\omega)\right|^2$$
$$=3\sum_{i}\Phi E\left[\left\{W_{\psi}\ddot{x}_{gm}''(a_j',b_i')\right\}^2\right].\left|H_1(\omega)\right|^2.H_2(\omega).\left|\hat{\psi}_{a_j',b_i'}(\omega)\right|^2 \tag{5.70}$$

The transfer function, $H_2(\omega)$, consists of a real component and an imaginary component, which can be easily derived from Equation (5.53) as follows:

$$H_2(\omega)=-\frac{\left(1-\omega^2\right)}{\left(1-\omega^2\right)^2+\left(2\zeta\omega\right)^2}+i\frac{2\zeta\omega}{\left(1-\omega^2\right)^2+\left(2\zeta\omega\right)^2} \tag{5.71}$$

However, only the real part of $H_2(\omega)$ must be considered on the right-hand side of Equation (5.70) because the left-hand side of the same equation is a real quantity. On denoting the real part $H_2(\omega)$ as β as follows,

$$\beta = -\frac{\left(1-\omega^2\right)}{\left(1-\omega^2\right)^2 + (2\zeta\omega)^2} \tag{5.72}$$

and using the following relation,

$$\int_{-\infty}^{\infty} \left|\hat{\psi}_{a_j',b_i'}(\omega)\right|^2 d\omega = 1 \tag{5.73}$$

and integrating Equation (5.70) over ω, one may get

$$\begin{aligned}&\sum_i E\left[W_\psi\Delta_1(a_j',b_i').W_\psi\Delta_0(a_j',b_i')\right]\\&= 3\sum_i E\left[\left\{W_\psi\, \ddot{x}_{gm}''(a_j',b_i')\right\}^2\right].\int_{-\infty}^{\infty}\beta.\Phi.\left|H_1(\omega)\right|^2\left|\hat{\psi}_{a_j',b_i'}(\omega)\right|^2 d\omega\end{aligned} \tag{5.74}$$

We are interested in finding out the instantaneous power spectral density function of the displacement response of the SDOF oscillator in consideration. To get this, it is necessary to first obtain the total instantaneous energy content of the system response over an interval Δb^* at a particular time point $t = b_i'$ where the energy over all frequency bands (i.e. over all j's) are summed up. Using Equations (5.64) and (5.74) to replace the expressions for $\{W_\psi\Delta_0^2(a_j',b_i')\}^2$ and $\{W_\psi\Delta_1(a_j',b_i').W_\psi\Delta_0(a_j',b_i')\}$, respectively, in the right-hand side of Equation (5.32), the expression for the total instantaneous energy content thus becomes

$$\begin{aligned}\sum_j E\left[\left\{X_i^j(\omega)\right\}^2\right] &= \sum_j \frac{K\Delta b^*}{a_j'} E\left[\left\{W_\psi\, \ddot{x}_{gm}''(a_j',b_i')\right\}^2\right]\int_{-\infty}^{\infty}\left|H_1(\omega)\right|^2\left|\hat{\psi}_{a_j',b_i'}(\omega)\right|^2 d\omega\\&+2\kappa.\left[3\sum_j \frac{K\Delta b^*}{a_j'} E\left[\left\{W_\psi\, \ddot{x}_{gm}''(a_j',b_i')\right\}^2\right].\int_{-\infty}^{\infty}\beta.\Phi\left|H_1(\omega)\right|^2.\left|\hat{\psi}_{a_j',b_i'}(\omega)\right|^2.d\omega\right]\end{aligned} \tag{5.75}$$

which may further be simplified to

$$\sum_j E\left[\left\{\Delta_i^j(\omega)\right\}^2\right] = \sum_j \frac{K\Delta b^*}{a_j'} E\left[\left\{W_\psi \ddot{x}_{gm}''(a_j', b_i')\right\}^2\right] \int_{-\infty}^{\infty} \left|H_1(\omega)\right|^2 \left|\hat{\psi}_{a_j', b_i'}(\omega)\right|^2 \left[1 + 6\kappa\beta\Phi\right] d\omega \tag{5.76}$$

The instantaneous power spectral density function of the displacement response of the oscillator (following guidelines laid out in Section 2.5) is obtained from Equation (5.76) by averaging the instantaneous total (summed up over all frequency bands) energy spectrum over Δb^* as written below:

$$S_\Delta(\omega) = \sum_j \frac{K}{a_j'} E\left[\left\{W_\psi \ddot{x}_{gm}''(a_j', b_i')\right\}^2\right] \left|H_1(\omega)\right|^2 \left|\hat{\psi}_{a_j', b_i'}(\omega)\right|^2 \left[1 + 6\kappa\beta\Phi\right] \tag{5.77}$$

The response of the oscillator may now be found from the instantaneous PSDF following the procedure described in Chapter 2. The moments of instantaneous PSDF of the nondimensional displacement term, $\Delta(T)$, must be evaluated to compute the statistics of the response process (i.e. the displacement term). The moments of this instantaneous output PSDF are used to obtain some essential statistical parameters like zero-crossing rates, bandwidth parameter, etc., which may be subsequently used to obtain the largest peak amplitude of the nonstationary stochastic random response process, $\Delta(T)$.

which may further be simplified to

[illegible]

Chapter 6

Wavelet-based probabilistic analysis

It was seen in Chapter 5 how system nonlinearity affects structural responses. The structure, in most of the cases, is supported on a soil medium that may behave nonlinearly, and this may affect the structural responses to a significant extent under the action of moderate to severe seismic ground motions. Several researchers have studied the influence of material nonlinearity of soil on the responses of a structure in the form of a nonlinear stress-strain relationship, transfer function analysis to obtain effective shear wave velocities in soil, a nonlinear load-displacement relationship, etc. [34–38]. The nonlinearity in the system (e.g. in case of slipping foundations) has been studied in the past [40] in the wavelet domain, extending the technique of equivalent linearization [38]. There is a general approach to model system nonlinearity using a hysteretic force-displacement relationship. However, in doing so to model large deformations, the area under the curve also gets larger, and hence a 'fat' hysteresis loop may be considered. In such cases, it is always worth modelling the effect of the hysteresis loop instead of representing the system stiffness by a backbone curve and an equivalent viscous damping mechanism (to represent dissipation of energy). In this chapter, the wavelet analytic technique is used to solve the system dynamics in the case of a tank supported on nonlinear soil. However, to keep things simpler, the force-deformation curve, in this case, is assumed to follow an elasto-perfectly-plastic (EPP) path, and subsequently a methodology for obtaining a nonstationary stochastic seismic response of a storage tank is described.

6.1 MODEL AND SOIL NONLINEARITY

In order to demonstrate the use of wavelets in solving the problems related to material nonlinearity, a structural model is considered similar to the one described in Chapter 3. The user may use the same concept for other structures as well. A rigid, right circular cylindrical tank is assumed to be supported on a thick circular foundation mat. This model, shown in Figure 6.1, is the same as that used in Chapter 3. However, the system now

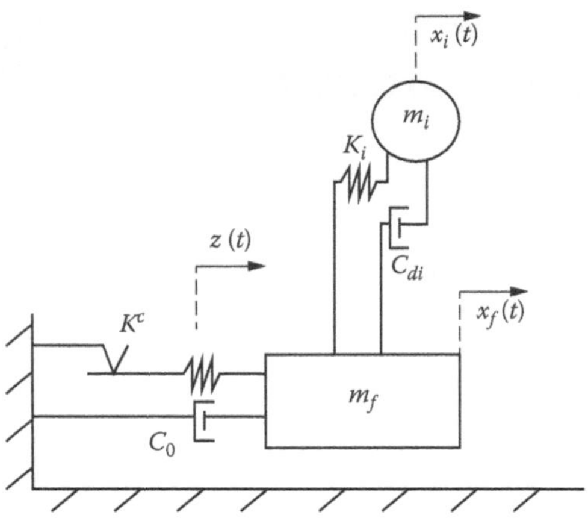

Figure 6.1 Tank–fluid–foundation model on soil idealized as a bilinear hysteretic EPP spring.

comprises only impulsive liquid mass and the supporting soil is considered to be nonlinear. It is known that many structures, subjected to seismic ground motions or another type of dynamic excitations, exhibit nonlinear behavior in the form of a hysteretic restoring force-displacement characteristic. In fact, hysteresis is encountered when there is relative sliding between substructures if the deformation amplitude exceeds a critical value. The hysteresis loops can be modelled in terms of a differential equation [38]. The basic concept in such modelling is to incorporate a differential equation into the general equations of motion of the system and derive an additional set of differential equations. These equations are then linearized and solved to obtain the system responses. This approach is adopted to formulate the equations of motion in the case of the tank–liquid–foundation vibration problem. The EPP model is used to represent the system nonlinearity. This EPP model is distinctly a bilinear model. The wavelet-based formulation discussed in previous sections is extended in this chapter to include pre-yield and post-yield phases. It may be noted here that the linear theory is valid in the two phases, as these phases are piece-wise linear. The spectral values are estimated later using probabilistic compatibility conditions. The liquid in the tank is vibrating in the same phase as that of the tank, so this is an impulsive mode of vibration for tank–liquid. The tank–liquid–foundation structure is resting on the surface of a nonlinear soil medium that is laterally flexible. The whole system is modelled by the combination of a mass-spring-dashpot system as shown in Figure 6.1. The impulsive mass is connected to the foundation mass through a spring and a damper. So, this spring and the damper represent the structural (material) properties of the tank. The supporting soil nonlinearity is considered here, assuming the soil to be connected to the superstructure by a hysteretic spring that is EPP in nature. At this point it is necessary to understand the working mechanism of the EPP hysteretic spring. The EPP nature means that the ratio of post-yield stiffness to pre-yield stiffness of the EPP spring is zero. The tank–foundation is actually connected to the soil by a combination of EPP hysteretic spring and a viscous damper. The source of excitation considered is the random, nonstationary ground excitation due to an earthquake. The deformation in the foundation of the tank due to seismic excitation induces a displacement in the attached spring until the yield level. Beyond the yield level, the associated friction element starts slipping. When the hysteretic loop unwinds, any change in foundation displacement causes a displacement in the EPP spring in the opposite direction. Due to this reason, the soil model considered here represents a bilinear hysteretic (i.e. nonlinear) system. The reader must be aware that the complication in analytical formulation may increase manifold if a complex nonlinear soil model is selected, and in most of these cases an analytical solution of the nonlinear differential equations may be difficult to obtain. However, a numerical modelling approach, especially based on the finite element method, may help in implementing a number of nonlinear soil models to simulate the true soil behaviour to a considerable extent. The application of a different type of nonlinear soil model is shown in Chapter 7. Wavelet-based system identification techniques for linear and nonlinear systems have been developed by Ghanem and Romeo [41, 42].

6.2 GENERAL EQUATIONS OF MOTION

In this chapter, the analytical formulation of a simple structure supported on a nonlinear soil-foundation system and subjected to a nonstationary random motion is developed as outlined in the work by Chatterjee and Basu [43]. A tank–fluid–foundation-soil system is assumed as a simple structure. The system is subjected to a lateral seismic excitation at its base. The nonstationary ground motion process, $\ddot{x}_g(t)$, causes a horizontal free-field ground displacement, $x_g(t)$, and shaking the superstructure as well. On considering the dynamic equilibrium of the tank–liquid–foundation system, the equation of motion is written in the following form, which was already seen in previous chapters:

$$m_i \ddot{x}_i(t) + C_{di} \ddot{x}_i(t) + K_i x_i(t) = -m_i(\ddot{x}_f(t) + \ddot{x}_g(t)) \tag{6.1}$$

In the above equation, the term m_i denotes the impulsive mass of the liquid participating in vibration. The terms $x_i(t)$ and $x_f(t)$ denote, respectively, the displacement of the liquid in the tank with respect to the foundation and the displacement of the foundation relative to the ground. The overdot represents acceleration. The terms C_{di} and K_i denote the viscous damping coefficient and static spring stiffness coefficient, respectively. These two linear coefficients represent the steel tank properties. The expressions of

C_{di} and K_i, though given in Chapter 3, are shown below once more for ready reference.

$$C_{di} = 2m_i\zeta_i\omega_i \tag{6.2}$$

$$K_i = m_i\omega_i^2 \tag{6.3}$$

The terms m_i, ζ_i and ω_i represent, respectively, the mass of the liquid in impulsive mode of vibration, the damping ratio of the impulsive mass and the first natural frequency of vibration of the tank–liquid. The expression of ω_i is given in Equation (3.25) in Chapter 3. It must be borne in mind that the natural frequency of vibration of the tank–liquid is the same as that of the tank itself as impulsive mode of vibration is considered. On substituting Equations (6.2) and (6.3), and then on dividing all terms by m_i, the following equation independent of impulsive mass is obtained:

$$\ddot{x}_i(t) + 2\zeta_i\omega_i\dot{x}_i(t) + \omega_i^2 x_i(t) + (\ddot{x}_f(t) + \ddot{x}_g(t)) = 0 \tag{6.4}$$

As it is seen that there are two degrees of freedom considered in the model (the displacements of the impulsive mass and the foundation), another equation has to be written to solve the system dynamics. The second equation comes from direct consideration of dynamic equilibrium of the total system (i.e. tank–liquid–foundation system) and may be written as

$$m_i\left\{\ddot{x}_i(t) + \ddot{x}_f(t) + \ddot{x}_g(t)\right\} + m_f\left\{\ddot{x}_f(t) + \ddot{x}_g(t)\right\} + C_0\,\ddot{x}_f(t) + K^c z(t) = 0 \tag{6.5}$$

In Equation (6.5), the readers may note three terms appearing – K^c, C_0 and $z(t)$. The term K^c is the static stiffness coefficient of the hysteretic spring, which attaches the foundation of the tank to the inelastic soil supporting the foundation. The other term, C_0, denotes the coefficient of the linear viscous damper, which is also placed between the tank–foundation and the inelastic soil. The expressions of K^c and C_0 are given as follows [33].

The expressions for viscous damper coefficient, C_0, and the static spring stiffness coefficient, K^c, are given below:

$$C_0 = \frac{R}{V_s}K^c\gamma_0 \tag{6.6}$$

and

$$K^c = \frac{8G_sR}{2-\nu_s} \tag{6.7}$$

In Equation (6.6), the term γ_0 appears, which is defined as follows:

$$\gamma_0 = 0.78 - 0.4\nu_s \tag{6.8}$$

The third term, $z(t)$, represents the displacement of the nonlinear hysteretic elasto-perfectly-plastic spring and is written as [38]

$$\dot{z}(t) = \dot{x}_f(t)[1 - F\{\dot{x}_f(t)\}F\{z(t) - x_y\} - F\{-\dot{x}_f(t)\}F\{-z(t) - x_y\}] \tag{6.9}$$

The soil beneath the tank–foundation system is assumed to have a yield displacement of x_y. In Equation (6.9), the term F represents unit step function (also called Heaviside step function). If the relative displacement of the EPP spring remains below the yield level or the hysteretic loop rewinds, the change in the displacement of the EPP spring becomes equal to the change in the displacement of the foundation. When the spring slips, there is no change in the hysteretic displacement of the spring even if the foundation undergoes deformation. In order to simplify Equation (6.5), all terms in the equation are divided by the common foundation mass term, m_f, and the terms C_0 and K^c are replaced with their expressions from Equations (6.6) and (6.7), respectively, to obtain the following equation:

$$\gamma_i\{\ddot{x}_i(t) + \ddot{x}_f(t) + \ddot{x}_g(t)\} + \ddot{x}_f(t) + \ddot{x}_g(t) + \frac{R}{V_s}\omega_{lf}^2\gamma_0\dot{x}_f(t) + \omega_{lf}^2 z(t) = 0 \tag{6.10}$$

It may be noted that the ratio of the impulsive liquid mass to the foundation mass is replaced with the term γ_i in Equation (6.10). The term ω_{lf} has been brought in to represent the lateral natural frequency of free vibration of the circular footing and is expressed as

$$\omega_{lf} = \sqrt{\frac{K^c}{m_f}} = \sqrt{\frac{8G_s R}{m_f(2 - \nu_s)}} \tag{6.11}$$

The term G_s denotes the shear modulus of the supporting soil and is given by Equation (4.17). The terms R and ν_s, respectively, represent the radius of the footing beneath the tank (which is assumed to be the same as that of the tank) and Poisson's ratio of the soil.

6.3 EQUATIONS BASED ON YIELD CONDITIONS

Equation (6.9) gives clear hints on how the solution could be obtained. A set of two distinct cases may arise depending upon the values of the hysteretic spring displacement ($z(t)$), the yield displacement (x_y) and the direction of the velocity of the foundation ($\dot{x}_f(t)$). One of the cases corresponds to the situation when the soil behaviour is linearly elastic, which is essentially the pre-yield case. The other one corresponds to the post-yield case, where the soil is in the plastic range. The conditions for these two cases are well defined and indeed depend on the direction of motion of the foundation and yield displacement. For the pre-yield case, the condition is that the foundation velocity is greater than zero when either the nonlinear EPP spring displacement is less than the yield displacement of the underlying soil medium, or the negative of the relative displacement of the EPP spring is greater than the yield displacement. In short, the condition for the pre-yield case may be written as $\dot{x}_f(t) > 0$ when $z(t) < x_y$, or $\dot{x}_f(t) < 0$ when $-z(t) < x_y$. For the post-yield case too, depending on whether the hysteretic spring displacement exceeds the yield displacement in either direction (+ or −), the condition may be concisely written as $\dot{x}_f(t) > 0$ when $z(t) > x_y$, or $\dot{x}_f(t) < 0$ when $-z(t) > x_y$. It is readily recognized that Equations (6.4) and (6.10) are valid in both pre-yield and post-yield cases because these equations are independent of yield conditions. However, depending upon the state of the yield, two separate equations of motion may be derived from Equation (6.9). On considering the conditions for the pre-yield case and post-yield case separately, the additional equations become

$$\dot{z}(t) = \dot{x}_f(t) \tag{6.12}$$

and

$$\dot{z}(t) = 0 \tag{6.13}$$

Thus, the equations that need to be solved representing the pre-yield situation are (6.4), (6.10) and (6.12), and those representing the post-yield case are (6.4), (6.10) and (6.13). Both sets of equations for the two yield cases are linear. In the next section, a wavelet-based solution for these cases will be formulated.

6.4 TRANSFER FUNCTIONS

In this section, two separate formulations will be developed – one for the pre-yield case and another for the post-yield case. It has been seen in earlier chapters that using a proper discretization scheme, the nonstationary random ground acceleration motion, $\ddot{x}_g(t)$, can be expressed in terms of wavelet functional (see Equation (2.11)). In the same way, the impulsive displacement response of the tank–liquid body (x_i), the displacement response of the foundation (x_f) and the relative displacement response of the EPP spring ($z(t)$) may be respectively written in terms of wavelet functionals as

$$x_i(t) = \sum_i \sum_j \frac{K\Delta b}{a_j} W_\psi x_i(a_j, b_i) \psi_{a_j,b_i}(t) \tag{6.14}$$

$$x_f(t) = \sum_i \sum_j \frac{K\Delta b}{a_j} W_\psi x_f(a_j, b_i) \psi_{a_j,b_i}(t) \tag{6.15}$$

$$z(t) = \sum_i \sum_j \frac{K\Delta b}{a_j} W_\psi z(a_j, b_i) \psi_{a_j,b_i}(t) \tag{6.16}$$

The expression for K is given in Equation (2.12). As the pre-yield case is first taken up, Equations (6.14) to (6.16) and Equation (2.12) may be substituted in the dynamic equations ((6.4), (6.10) and (6.12)) of the system to obtain the following wavelet-based equations:

$$\begin{aligned}
&\sum_i \sum_j \frac{K\Delta b}{a_j} W_\psi x_i(a_j, b_i) \ddot{\psi}_{a_j,b_i}(t) + 2\zeta_i \omega_i \sum_i \sum_j \frac{K\Delta b}{a_j} W_\psi x_i(a_j, b_i) \dot{\psi}_{a_j,b_i}(t) \\
&+ \omega_i^2 \sum_i \sum_j \frac{K\Delta b}{a_j} W_\psi x_i(a_j, b_i) \psi_{a_j,b_i}(t) \\
&+ \left(\sum_i \sum_j \frac{K\Delta b}{a_j} W_\psi x_f(a_j, b_i) \ddot{\psi}_{a_j,b_i}(t) + \sum_i \sum_j \frac{K\Delta b}{a_j} W_\psi \ddot{x}_g(a_j, b_i) \psi_{a_j,b_i}(t) \right) = 0
\end{aligned} \tag{6.17}$$

$$\begin{aligned}
&\gamma_i \left(\sum_i \sum_j \frac{K\Delta b}{a_j} W_\psi x_i(a_j, b_i) \ddot{\psi}_{a_j,b_i}(t) + \sum_i \sum_j \frac{K\Delta b}{a_j} W_\psi x_f(a_j, b_i) \ddot{\psi}_{a_j,b_i}(t) \right. \\
&\left. + \sum_i \sum_j \frac{K\Delta b}{a_j} W_\psi \ddot{x}_g(a_j, b_i) \psi_{a_j,b_i}(t) \right) + \sum_i \sum_j \frac{K\Delta b}{a_j} W_\psi x_f(a_j, b_i) \ddot{\psi}_{a_j,b_i}(t) \\
&+ \sum_i \sum_j \frac{K\Delta b}{a_j} W_\psi \ddot{x}_g(a_j, b_i) \psi_{a_j,b_i}(t) + \frac{R}{V_s} \omega_{lf}^2 \gamma_0 \sum_i \sum_j \frac{K\Delta b}{a_j} W_\psi x_f(a_j, b_i) \dot{\psi}_{a_j,b_i}(t) \\
&+ \omega_{lf}^2 \sum_i \sum_j \frac{K\Delta b}{a_j} W_\psi z(a_j, b_i) \psi_{a_j,b_i}(t) = 0
\end{aligned} \tag{6.18}$$

$$\sum_i \sum_j \frac{K\Delta b}{a_j} W_\psi z(a_j, b_i) \dot{\psi}_{a_j,b_i}(t) = \sum_i \sum_j \frac{K\Delta b}{a_j} W_\psi x_f(a_j, b_i) \dot{\psi}_{a_j,b_i}(t) \tag{6.19}$$

The above system equations in the wavelet domain are written following the same procedure as described in Chapter 2. Now, after cancelling the common term $K\Delta b$ throughout, rearranging the terms properly and thereafter taking Fourier transform of all terms of Equations (6.17), (6.18) and (6.19), one may easily obtain the following equations:

$$\begin{aligned}
&\sum_i \sum_j \frac{1}{a_j} W_\psi x_i(a_j, b_i)\left\{-\omega^2 \hat{\psi}_{a_j,b_i}(\omega)\right\} + 2i\zeta_i\omega_i\omega \sum_i \sum_j \frac{1}{a_j} W_\psi x_i(a_j, b_i)\hat{\psi}_{a_j,b_i}(\omega) \\
&+ \omega_i^2 \sum_i \sum_j \frac{1}{a_j} W_\psi x_i(a_j, b_i)\hat{\psi}_{a_j,b_i}(\omega) + \left(\sum_i \sum_j \frac{1}{a_j} W_\psi x_f(a_j, b_i)(-\omega^2)\hat{\psi}_{a_j,b_i}(\omega) \right. \\
&\left. + \sum_i \sum_j \frac{1}{a_j} W_\psi \ddot{x}_g(a_j, b_i)\hat{\psi}_{a_j,b_i}(\omega) \right) = 0
\end{aligned} \tag{6.20}$$

$$\begin{aligned}
&\gamma_i \left(\sum_i \sum_j \frac{1}{a_j} W_\psi x_i(a_j, b_i)\hat{\psi}_{a_j,b_i}(\omega)(-\omega^2) + \sum_i \sum_j \frac{1}{a_j} W_\psi x_f(a_j, b_i)\hat{\psi}_{a_j,b_i}(\omega)(-\omega^2) \right. \\
&\left. + \sum_i \sum_j \frac{1}{a_j} W_\psi \ddot{x}_g(a_j, b_i)\hat{\psi}_{a_j,b_i}(\omega) \right) + \sum_i \sum_j \frac{1}{a_j} W_\psi x_f(a_j, b_i)\hat{\psi}_{a_j,b_i}(\omega)(-\omega^2) \\
&+ \sum_i \sum_j \frac{1}{a_j} W_\psi \ddot{x}_g(a_j, b_i)\hat{\psi}_{a_j,b_i}(\omega) \\
&+ \frac{R}{V_s}\omega_{lf}^2 \gamma_0 \sum_i \sum_j \frac{1}{a_j} W_\psi x_f(a_j, b_i)\hat{\psi}_{a_j,b_i}(\omega)(i\omega) = 0
\end{aligned} \tag{6.21}$$

$$\sum_i \sum_j \frac{1}{a_j} W_\psi z(a_j, b_i)\hat{\psi}_{a_j,b_i}(\omega)(i\omega) = \sum_i \sum_j \frac{1}{a_j} W_\psi x_f(a_j, b_i)\hat{\psi}_{a_j,b_i}(\omega)(i\omega) \tag{6.22}$$

At this point, Equation (2.40) should be recalled, which states the orthogonality property of the wavelet basis function, and also Equation (4.31) has to be used, which gives the relation between ground displacement and ground acceleration through wavelet coefficients. Thus, on multiplying all terms in

Equation (6.20) by $\hat{\psi}^*_{a_k,b_l}(\omega)$, and subsequently on using Equations (2.40) and (4.31) and collecting terms properly, the following equation may be written:

$$\sum_i W_\psi x_i(a_j,b_i)\hat{\psi}_{a_j,b_i}(\omega)\left(\omega_i^2-\omega^2+2i\zeta_i\omega_i\omega\right)$$
$$+\sum_i W_\psi x_f(a_j,b_i)\hat{\psi}_{a_j,b_i}(\omega)\left(-\omega^2\right)=-\sum_i W_\psi \ddot{x}_g(a_j,b_i)\hat{\psi}_{a_j,b_i}(\omega) \tag{6.23}$$

Similar operations could be performed on Equations (6.21) and (6.22) to obtain the following two wavelet domain equations:

$$\sum_i W_\psi x_i(a_j,b_i)\hat{\psi}_{a_j,b_i}(\omega)\left(-\gamma_i\omega_i^2\right)$$
$$+\sum_i W_\psi x_f(a_j,b_i)\hat{\psi}_{a_j,b_i}(\omega)\left\{-\omega^2(1+\gamma_i)+i\frac{R}{V_s}\gamma_0\omega\omega_{lf}^2\right\} \tag{6.24}$$
$$+\sum_i W_\psi z(a_j,b_i)\hat{\psi}_{a_j,b_i}(\omega)\left(\omega_{lf}^2\right)=-\sum_i W_\psi \ddot{x}_g(a_j,b_i)\hat{\psi}_{a_j,b_i}(\omega)(1+\gamma_i)$$

$$\sum_i W_\psi z(a_j,b_i)\hat{\psi}_{a_j,b_i}(a_j,b_i)(\omega)=\sum_i W_\psi x_f(a_j,b_i)\hat{\psi}_{a_j,b_i}(\omega) \tag{6.25}$$

In order to obtain the wavelet-based response in terms of displacements of impulsive liquid mass of the tank and the displacements of the foundation of the tank, Equations (6.23) to (6.25) must be solved. After doing some simple algebraic manipulation, the user should obtain the two expressions as shown below – one for the impulsive displacement response and the next for the foundation displacement response.

$$\sum_i W_\psi x_i(a_j,b_i)\hat{\psi}_{a_j,b_i}(\omega)=\tau_i^{pre}(\omega)\sum_i W_\psi \ddot{x}_g(a_j,b_i)\hat{\psi}_{a_j,b_i}(\omega) \tag{6.26}$$

$$\sum_i W_\psi x_f(a_j,b_i)\hat{\psi}_{a_j,b_i}(\omega)=\tau_f^{pre}(\omega)\sum_i W_\psi \ddot{x}_g(a_j,b_i)\hat{\psi}_{a_j,b_i}(\omega) \tag{6.27}$$

It is seen from the above equations that the displacement responses are tied up to the ground motion through complex-valued frequency-dependent transfer functions ($\tau_i^{pre}(\omega)$ for impulsive displacement response and $\tau_f^{pre}(\omega)$ for foundation displacement response). The main objective of wavelet-based analysis is to finally obtain a relation between the wavelet coefficients of response terms and the wavelet coefficients of the input excitation through transfer functions. These transfer functions will be required to obtain expected largest peak values after some statistical evaluations. This has also been seen in previous chapters. For the present case, these transfer functions for the pre-yield case may be elaborated as follows:

$$\tau_i^{pre}(\omega) = \frac{\omega_{lf}^2 + i\left(\gamma_0 a_0 \omega_{lf}^2\right)}{[C]+i[D]} \tag{6.28}$$

$$\tau_f^{pre}(\omega) = \frac{\left\{(1+\gamma_i)\omega_i^2 - \omega^2\right\} + i\left\{2\zeta_i \omega\omega_i (1+\gamma_i)\right\}}{[C]+i[D]} \tag{6.29}$$

The terms C and D are given as

$$C = \left(\omega_i^2 - \omega^2\right)\left\{\omega_{lf}^2 - (1+\gamma_i)\omega^2\right\} - 2\zeta_i \omega\omega_i a_0 \omega_{lf}^2 \gamma_0 - \gamma_i \omega^4 \tag{6.30}$$

$$D = 2\zeta_i \omega\omega_i \left\{\omega_{lf}^2 - (1+\gamma_i)\omega^2\right\} + \left(\omega_i^2 - \omega^2\right) a_0 \omega_{lf}^2 \gamma_0 \tag{6.31}$$

The term a_0, appearing in the above expressions, represents a dimensionless frequency parameter, as given by Equation (4.18). With this, the formulation of equations, and obtaining transfer functions corresponding to the pre-yield case, now has become complete. The next objective is to obtain similar expressions for the post-yield case.

For the post-yield case, the dynamic equilibrium equations for the tank–liquid system and the tank–liquid–foundation system remain the same, so Equations (6.4) and (6.10) may be reused for the post-yield case along with Equation (6.13), which is specific to the yield case (post-yield in this case). The wavelet domain representation of Equation (6.13) may be obtained following a procedure similar to that described while formulating equations in the pre-yield case, and is given below:

$$\sum_i W_\psi z(a_j, b_i)\hat{\psi}_{a_j,b_i}(\omega) = 0 \tag{6.32}$$

Thus, for the post-yield case, Equations (6.23), (6.24) and (6.32) must be solved to obtain the transfer functions $\tau_i^{post}(\omega)$ and $\tau_f^{post}(\omega)$, relating the wavelet coefficients of impulsive liquid displacements and foundation displacements, respectively, to the wavelet coefficients of seismic ground acceleration. These transfer functions are written below:

$$\tau_i^{post}(\omega) = -\frac{i\left(\gamma_0 a_0 \omega_{lf}^2\right)}{[E]+i[F]} \tag{6.33}$$

$$\tau_f^{post}(\omega) = -\frac{\left\{(1+\gamma_i)\omega_i^2 - \omega^2\right\} + i\left\{2\zeta_i \omega\omega_i (1+\gamma_i)\right\}}{[E]+i[F]} \tag{6.34}$$

The terms E and F in the expressions of post-yield transfer functions are longer terms that are shown below separately.

$$E=\left(\omega_i^2-\omega^2\right)\left\{\omega_{lf}^2-(1+\gamma_i)\omega_i^2\right\}-2\zeta_i\omega\omega_i\gamma_0 a_0\omega_{lf}^2-\gamma_i^4 \tag{6.35}$$

$$F=2\zeta_i\omega_i\omega\left\{\omega_{lf}^2-\omega^2(1+\gamma_i)\right\}+\left(\omega_i^2-\omega^2\right)\gamma_0 a_0\omega_{lf}^2 \tag{6.36}$$

Once these transfer functions are known, the impulsive displacement response of the tank may be derived using wavelet-based relations for the pre-yield and post-yield cases.

6.5 RESPONSE OF THE STRUCTURE

In this section, the response of the tank in terms of the displacement of the tank–liquid undergoing impulsive mode of vibration will be obtained in the wavelet domain. The displacement response, as will be seen, depends on the probability of yield crossing. Following Equation (2.49), the instantaneous mean square energy response for the jth energy band corresponding to the displacement of the impulsive mode of the tank–liquid may be written as shown below:

$$E\left[\left|X_i^j(\omega)\right|^2\right]=\frac{K\Delta b}{a_j}E\left[\left|W_\psi\,\ddot{x}_g(a_j,b_i)\right|^2\right]\int_{-\infty}^{\infty}|H(\omega)|^2\left|\hat{\psi}_{a_j,b_i}(\omega)\right|^2 d\omega \tag{6.37}$$

In this equation of instantaneous mean square energy response of impulsive tank–liquid displacement contained in the jth frequency band, $H(\omega)$ denotes the corresponding transfer function, which is dependent on the state of the yield displacement – when it is the pre-yield state, $H(\omega)=\tau_i^{pre}(\omega)$, and when it is the post-yield state, $H(\omega)=\tau_i^{post}(\omega)$. Now, the instantaneous power spectral density function (PSDF) of the displacement response of the tank–liquid may be computed based on the expected energy summed up over all frequency bands and subsequently averaging it over the small time interval of Δb. Thus, following Equation (2.50), the PSDF at a particular instant in time may be written as

$$S_{x_i}(\omega)=\sum_j\frac{E[|X_i^j(\omega)|^2]}{\Delta b}=\sum_j\frac{K}{a_j}E\left[\left|W_\psi\,\ddot{x}_g(a_j,b_i)\right|^2\right]|H(\omega)|^2\left|\hat{\psi}_{a_j,b_i}(\omega)\right|^2 \tag{6.38}$$

It is seen from the standard approach as described in Chapter 2 that the next step would be to evaluate the instantaneous root mean square

(RMS) values of the displacement responses of the structure, which in this case is the impulsive tank–liquid displacement. To do so, only zeroth moments of instantaneous PSDF should be computed first. The calculation of zeroth moment depends heavily on the transfer function $H(\omega)$, which in turn is dependent on the yield condition (whether pre-yield or post-yield). However, at any specific time point, the yield-crossing phenomenon of the hysteretic spring is random in nature, which means the response of the system would also be a stochastic phenomenon. This indicates that the probability of yield crossing has to be calculated in order to compute the correct response. Based on the probability of occurrence of the spring in the pre-yield or post-yield state, the instantaneous power spectral density function must be evaluated.

We have the following four yield conditions as explained already in Section 6.3: (1) $\dot{x}_f > 0,\ z < x_y$, (2) $\dot{x}_f < 0,\ -z < x_y$, (3) $\dot{x}_f > 0,\ z > x_y$ and (4) $\dot{x}_f \langle 0, -z \rangle x_y$. The first two represent the pre-yield case, and the next two represent the post-yield case. The probabilities that the spring displacement would satisfy the yield conditions 1, 2, 3 and 4 are denoted by p_1, p_2, p_3 and p_4, respectively. The zeroth moment of the instantaneous PSDF of the structural displacement response depends on these probabilities and may be written as

$$
\begin{aligned}
m_{0|t=b_i} = {} & p_1\left\{\sum_j K'E\left[\left|W\ddot{x}_g(a_j,b_i)\right|^2\right]\right\}\left|I_{o,j}\right|_{|\dot{x}_f>0,\ z<x_y} \\
& + p_2\left\{\sum_j K'E\left[\left|W\ddot{x}_g(a_j,b_i)\right|^2\right]\right\}\left|I_{o,j}\right|_{|\dot{x}_f<0,\ -z<x_y} \\
& + p_3\left\{\sum_j K'E\left[\left|W\ddot{x}_g(a_j,b_i)\right|^2\right]\right\}\left|I'_{0,j}\right|_{|\dot{x}_f>0,\ z>x_y} \\
& + p_4\left\{\sum_j K'E\left[\left|W\ddot{x}_g(a_j,b_i)\right|^2\right]\right\}\left|I'_{0,j}\right|_{|\dot{x}_f<0,\ -z>x_y}
\end{aligned}
\tag{6.39}
$$

The readers must have noticed that the expression for zeroth moment of instantaneous PSDF contains terms $I_{o,j}$ and $I'_{0,j}$ for pre-yield and post-yield cases, respectively. These terms are similar to the ones described earlier in previous chapters and are defined below:

$$
I_{o,j} = \int_{\frac{\pi}{a_j}}^{\frac{\sigma\pi}{a_j}} \frac{\left(\omega_{lf}^2\right)^2 + \left(\gamma_0 a_0 \omega_{lf}^2\right)^2}{E^2 + F^2}
\tag{6.40}
$$

Thus, it is seen that the computation of instantaneous PSDF requires the values of the probabilities beforehand. The method of calculation of these probabilities is explained in the next section.

6.6 PROBABILITY EVALUATION

The normal or Gaussian distribution is well known and defined as

$$f(x,\mu,\sigma)=\frac{1}{\sigma\sqrt{2\pi}}e^{-\frac{(x-\mu)^2}{2\sigma^2}} \tag{6.41}$$

The terms σ, μ and σ^2, respectively, denote the standard deviation, the mean and the variance of the distribution. We will use the Gaussian distribution curve to evaluate the probabilities. For this, we assume that the hysteretic spring displacement, z, has a zero mean, mixed Gaussian distribution locally. The distribution is assumed to be truncated because the displacement of the spring (z) in the pre-yield case remains within the range $-x_y \le z \le x_y$. The hysteretic spring displacement term, z, being a random variable, may assume any continuous value in the range of $-x_y \le z \le x_y$ in the pre-yield case or a specific value (i.e. either $-x_y$ or x_y) in the post-yield case. Hence, this Gaussian distribution is also assumed to be mixed (continuous-discrete). The zero mean, locally truncated, mixed Gaussian distribution of z may be defined completely by the following equation:

$$p(z)=\frac{1}{a\sqrt{2\pi}}e^{-\left(\frac{z^2}{2a^2}\right)}; \quad -x_y < z < x_y \tag{6.42}$$

$$p(z)=p'; \quad z=x_y \text{ or } z=-x_y \tag{6.43}$$

This mixed Gaussian distribution, with a being a parameter, is shown in Figure 6.2. As this is a nonstationary case, the distribution is local at any time instant t. The parameter σ also varies with time. The probability $\tilde{p}$ that the spring displacement z lies between $-x_y$ and x_y (i.e. $-x\_y < z < x\_y$) to represent the pre-yield situation can be obtained from calculating the area under the Gaussian distribution curve between the same limits of z as follows:

$$\tilde{p}=\frac{1}{a\sqrt{2\pi}}\int_{-x_y}^{x_y} e^{\left(-\frac{z^2}{2a^2}\right)}dz \tag{6.44}$$

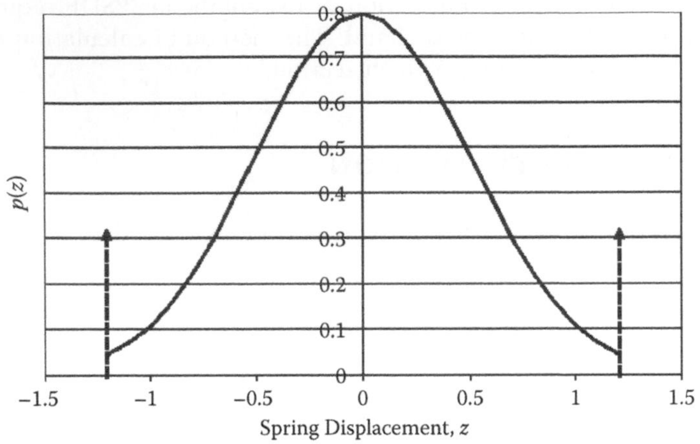

Figure 6.2 Mixed Gaussian distribution.

Further simplification of Equation (6.44) may be done using a substitution, $z = va$, in the equation as

$$\tilde{p} = \frac{1}{\sqrt{2\pi}} \int_{-\frac{x_y}{a}}^{\frac{x_y}{a}} e^{\left(-0.5v^2\right)} dz \tag{6.45}$$

The probability p' that the event remains in the post-yield state (i.e. equal to either $-x_y$ or x_y) may be calculated from the fact that the total probability of the variable z lying in both pre-yield and post-yield states (i.e. $-x_y \le z \le x_y$) is unity. Thus, $2p' + \tilde{p} = 1$, from which we get

$$p' = \frac{1-\tilde{p}}{2} \tag{6.46}$$

The probability that the foundation displacement x_f occurs in either direction (i.e. $x_f < 0$ or $x_f > 0$) is equal to 0.5. If x_f is assumed to be locally Gaussian, the expressions for the probabilities p_1 $(= \Pr[\dot{x}_f > 0,\ z < x_y] = \Pr[\dot{x}_f > 0] - \Pr[\dot{x}_f > 0, z \ge x_y])$ and p_2 $(= \Pr[\dot{x}_f < 0,\ -z < x_y] = \Pr[\dot{x}_f > 0] - \Pr[\dot{x}_f < 0, -z \ge x_y])$ in the pre-yield case are given as

$$p_1 = p_2 = \frac{\tilde{p}}{2} \tag{6.47}$$

Similarly, the expressions for probabilities p_3 $(= \Pr[\dot{x}_f > 0, z \geq x_y] = \Pr[\dot{x}_f > 0 | z > x_y].\Pr[z \geq x_y])$ and p_4 $(= \Pr[\dot{x}_f < 0, -z \geq x_y] = \Pr[\dot{x}_f < 0 | -z > x_y].\Pr[-z \geq x_y])$ in the post-yield case are defined as

$$p_3 = p_4 = \frac{1-\tilde{p}}{2} \tag{6.48}$$

If it is also assumed that the permanent drifting (or the yield) of the foundation is reasonably low in the pre-yield situation, the expected mean square value of the hysteretic spring displacement (relative), $z(t)$, remains the same as the foundation displacement (relative), $x_f(t)$. However, the expected mean square value of $z(t)$ in the post-yield state remains constant and is equal to x_y^2. The instantaneous variance of hysteretic spring displacement may be calculated using wavelet functionals considering the chances of occurrence of $z(t)$ in the pre-yield and post-yield cases and using Equations (6.47) and (6.48). Following Chapter 2, the expression of this variance, $m_0^z|_{t=b_i}$, for the pre-yield case is given below:

$$m_0^z|_{t=b_i} = \tilde{p}\left[\sum_j K' E\{W\ddot{x}_g(a_j,b_i)^2\}\right].\left[I_{0,j}^z\right] + (1-\tilde{p})x_y^2 \tag{6.49}$$

The term $I_{0,j}^z$ for the relative hysteretic spring displacement is the same as the relative foundation displacement in the pre-yield case and may be expressed as

$$I_{0,j}^z = \int_{\pi/a_j}^{\sigma\pi/a_j} \frac{\{(1+\gamma_i)\omega_i^2 - \omega^2\}^2 + \{2\zeta_i\omega\omega_i(1+\gamma_i)\}^2}{C^2 + D^2} d\omega \tag{6.50}$$

where the expressions for C and D are given in Equations (6.30) and (6.31), respectively. It may be seen from Equations (6.39), (6.47), (6.48) and (6.49) that the evaluation of probabilities p_1 through p_4 and the calculation of instantaneous moment depend upon the correct evaluation of the term $\tilde{p}$. This is done as follows.

The variance of $z(t)$, i.e. $m_0^z|_{t=b_i}$, is calculated as per Equation (6.49). However, the same variance may be calculated alternatively using the probability distribution from the following equation:

$$m_0^z|_{t=b_i} = \frac{a^2}{\sqrt{2\pi}} \int_{-x_y/a}^{x_y/a} e^{-0.5u^2} u^2 du + (1-\tilde{p})x_y^2 \tag{6.51}$$

For a given value of yield displacement x_y, the value of a can be calculated once an initial value of $\frac{x_y}{a}$ is selected. With this value of x_y, the value of $\tilde{p}$ can be calculated from the integral given in Equation (6.45) and may be subsequently used to obtain the values of the probabilities p_1, p_2, p_3 and p_4 from Equations (6.47) and (6.48). Once these probabilities are known, $m_0^{\dot{z}}|_{t=b_i}$ may be evaluated from Equation (6.49). The value of $\tilde{p}$ obtained from Equation (6.45) may be considered to be the correct one only when the value of $m_0^{\dot{z}}|_{t=b_i}$ obtained so far (from Equation (6.49)) matches with the value of $m_0^{\dot{z}}|_{t=b_i}$ obtained from Equation (6.51). If the equality of $m_0^{\dot{z}}|_{t=b_i}$ obtained from Equations (6.49) and (6.51) is not established, the whole procedure has to be iterated again. However, on obtaining the correct value of $\tilde{p}$, the instantaneous moments for the displacement of the impulsive liquid mass in the tank may be obtained from Equation (6.39).

6.7 VALIDATION AND RESULTS

Let us present the concept explained so far using the same tank–liquid–foundation model subjected to seismic ground motion. The tank–liquid model is assumed to be slender with a large value of height–radius ratio (3 in this case), which allows the liquid in the tank to vibrate mostly in impulsive mode. The tank wall is quite thin with a wall thickness-to-tank radius ratio (i.e. h/R ratio) being confined to 0.001. Once this ratio is known, the appropriate value of C_i in Equation (3.25) may be chosen from Table 3.1. The material properties of the steel tank are as follows. Young's modulus = 2.1×10^{11} N/m^2, Poisson's ratio = 0.30, mass density = 7850 kg/m^3 and damping ratio = 5%. The ratio of the density of water (the tank–liquid) is assumed as 1000 kg/m^3. The Poisson's ratio of the soil supporting the tank–foundation system is kept constant at 0.33, and the lateral natural frequency of the tank–foundation (ω_{lf}) is assumed to be 10 Hz (or ~62.83 rad/s). The impulsive liquid mass is taken as eight times heavier than the mass of the foundation (i.e. γ_i = 8). The ground motion process considered here corresponds to the 1989 Loma Prieta earthquake at the Dumbarton Bridge site near Coyote Hills.

It is assumed that a right circular cylindrical tank with material properties as described above is undergoing seismic excitation as per the time history shown in Figure 1.2. The tank has a height of 6 m and is supported on a soil with shear wave propagation velocity of 200 m/s and yield displacement of 0.0025 m. The wavelet-based nonlinear stochastic response of the tank is obtained in terms of instantaneous root mean square of the displacement of the impulsive mode of the tank–liquid system as per the formulation laid above. Figure 6.3 represents comparison of responses between the curves obtained from the wavelet-based analysis and exact time history simulation (using the Runge-Kutta fourth-order method).

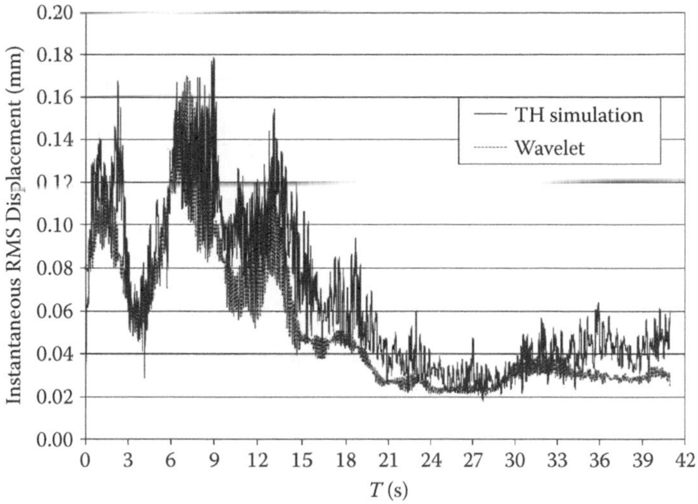

Figure 6.3 Instantaneous RMS displacement response for a tank with height of 6 m.

The comparisons are repeated for tanks with heights of 4 and 10 m, while other properties of the tank and the underlying supporting soil remain the same. These comparisons are shown in Figures 6.4 and 6.5.

To look into more detail about the displacement response of the tank, the tank is subjected to an accelerogram that is three times more magnified

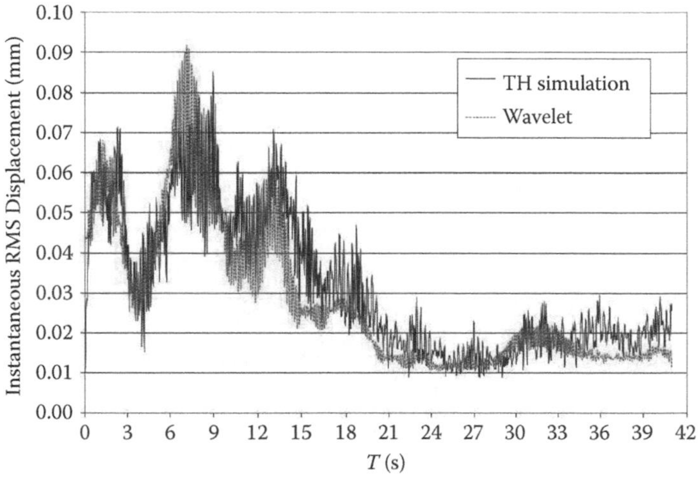

Figure 6.4 Instantaneous RMS displacement response for a tank with height of 4 m.

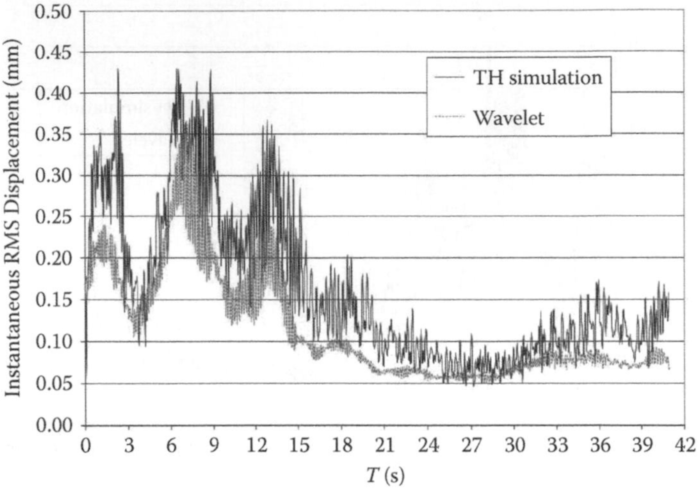

Figure 6.5 Instantaneous RMS displacement response for a tank with height of 10 m.

than the previous ones. This would reveal clearly the effect of strong ground motion nonstationarities on the structural responses. It is clearly seen from Figure 6.6 that the instantaneous RMS displacement response of the tank increases with an increase in tank height, so the more the tank height, the more pronounced becomes the effect of ground motions. As the yield displacement is kept the same when compared to the tank considered

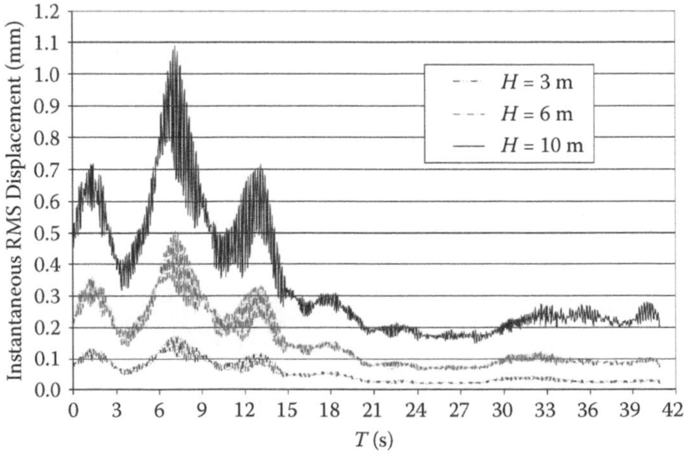

Figure 6.6 Instantaneous RMS displacement response for tanks with different heights.

to generate responses shown in Figures 6.3 to 6.5, the amplified ground motion pushes the tank displacement into the nonlinear range quickly and for a longer time. However, in Figure 6.6, the peaks seem to occur at the same place in the time domain for all heights of the tank, which is quite obvious, as the accelerogram, though amplified, does not have any change in its frequency content.

Chapter 7

General applications

In previous chapters we have seen how the wavelet-based analytical technique can be used to formulate equations of motion in the wavelet domain in steps, and subsequently how the system equations (for both linear and nonlinear systems) can be solved to obtain the expected largest peak responses of the system. In all cases, we have seen that some common objectives have been met in order to solve the dynamic equilibrium in the wavelet domain. In this chapter, we will discuss some interesting applications of wavelets in different fields of civil engineering. However, before starting the details of these applications, it would be wise to briefly discuss once more the general properties, strengths as well as shortcomings, of wavelets that the users must be aware of before indulging in wavelet-based analysis.

We have seen that Fourier transform can give us information about the frequency content of a signal, but the time information gets lost, and this renders Fourier transform less usable in case the signal corresponds to a nonstationary random process. The wavelet transforms enable us to obtain orthonormal basis expansions of a signal using mother wavelets that have a good time–frequency localization property. The wavelet transform is based on octave band decomposition in the time–frequency plane in which higher frequencies are localized well in time. However, as frequency increases, uncertainty in frequency localization also increases. In wavelet transform, the large-scale features are well resolved in the time domain with greater uncertainty in their location. The small-scale features are well resolved in the time domain with greater uncertainty with their frequency content. This phenomenon is due to the limitation explained by Heisenberg's uncertainty principle. The choice of a wavelet depends on many factors, including the purpose of the analysis and the type of the signal used, and hence this choice cannot be made arbitrarily. The function (t) is chosen with unit energy (i.e. integration ($|(t)|^2 = 1$) such that it has compact support, indicating that the decay would be fast, and hence localization in the spatial domain would be possible. The mother wavelet is also chosen to ensure that it has a proper admissibility condition; i.e. it has zero mean (i.e. $\int_{-\infty}^{\infty}(t)dt=0$), and also the higher moments may be zero as well, as shown below:

$$\int_{-\infty}^{\infty} t^k(t)dt = 0; \quad k = 0,1,\ldots,N-1 \tag{7.1}$$

This zero mean property ensures that the function has a wave-like shape. Thus, several functions satisfying these two conditions may be chosen. The wavelet transform is an energy-preserving transformation. In order to take the wavelet transform of a sampled signal, discretization of scale and location parameters (or time and frequency parameters) must be done, and this is called discrete wavelet transform, which is already explained in Chapter 2.

7.1 B-WIM NOR SIGNAL ANALYSIS

Weighing-in-motion (WIM) devices are designed to capture and record axle weights and gross vehicle weights as vehicles drive over a measurement site. Unlike static scales, WIM systems are capable of measuring vehicles travelling at a reduced or normal traffic speed and do not require the vehicle to come to a stop. This makes the weighing process more efficient and, in the case of commercial vehicles, allows for trucks under the weight limit to bypass static scales or inspection. Bridge weigh-in-motion (B-WIM) is a process to determine the axle and gross vehicle weights of vehicles travelling at highway speeds on instrumented bridges. In this case, strain transducers are placed underneath the bridge for axle detection and nothing-on-the-road (NOR). The signals, measuring bridge vibrations due to the passage of traffic over the bridge, are recorded in these transducers and later analysed by wavelets to extract detailed information that otherwise would not have been possible. If $_{s,\tau}(t)$ is chosen as the mother wavelet, the continuous wavelet transform may be defined as

$$Wx(s,\tau) = \int x(t)_{s,\tau}(t)dt \tag{7.2}$$

However, the continuous wavelet transform cannot be computed in a practical sense, as the variables involved in the definition of continuous wavelet transform are continuous. Hence, the transformation needs discretization in such a way that at lower frequencies the sampling rate may be decreased, thereby significantly saving computation time. An example of the B-WIM problem is given below. A numerical B-WIM model is described and realistic values of the parameters of the model are given. Subsequently, the numerically obtained strain signals are analysed by appropriate wavelets to reveal more information. Additionally, experimentally obtained strain signals are also analysed to obtain similar information on axle spacing, vehicle velocity, etc. Chatterjee et al. [44] showed efficient use of wavelets in signal processing with respect to B-WIM to extract vehicle information.

7.1.1 Bridge and vehicle model

A two-dimensional numerical walking beam model is used to generate strain signals corresponding to a biaxial tandem travelling at a speed of approximately 75 km/h over a 15 m long simply supported beam. The vehicle is assumed to have a total mass of about 12 440 kg. This mass is assumed to be located at mid-way between the two axles. The bridge is assumed to have a mass of 28 125 kg/m, considering the length of the bridge to be the same as the beam (i.e. 15 m). The model of the vehicle is shown in Figure 7.1. The distance between the axles is 1.5 m. The model has vertical translation and a rotation (hence, it is a two-degree-of-freedom (2-DOF) model). In the model, the terms K and C denote the stiffness coefficients and dampers, respectively. The mass is denoted by m and the velocity by $V(x, t)$.

The moment of inertia of the vehicle is 45 000 kgm^2, whereas the second moment of inertia of the cross section of the bridge is 0.527 m^4. The vehicle suspension is modeled by two spring-dashpot combinations that have spring stiffnesses of 350 kN/m and damping coefficients of 7 kN-s/m. The modulus of elasticity of the beam is 35×10^9 N/m^2. The second-order equation of dynamic equilibrium of the model may be solved to obtain the interaction force between the bridge and the vehicle axles at a particular time step and may be used in the next time step to update the beam (or bridge) deflection [44]. The strains are calculated at one-quarter and three-quarter points (i.e. $0.25L$ and $0.75L$) on the beam (L = 15 m). The strains are sampled at 6300 Hz. The amplitude of vibration response obtained from the numerical solution of the system equation is quite high, and hence it is difficult to identify the axle passage

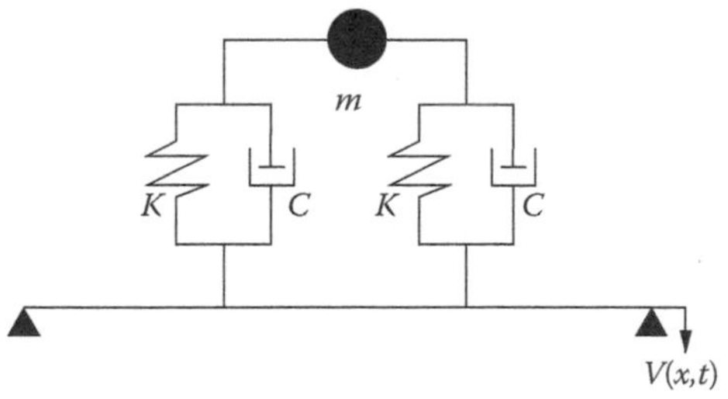

Figure 7.1 Model of a vehicle.

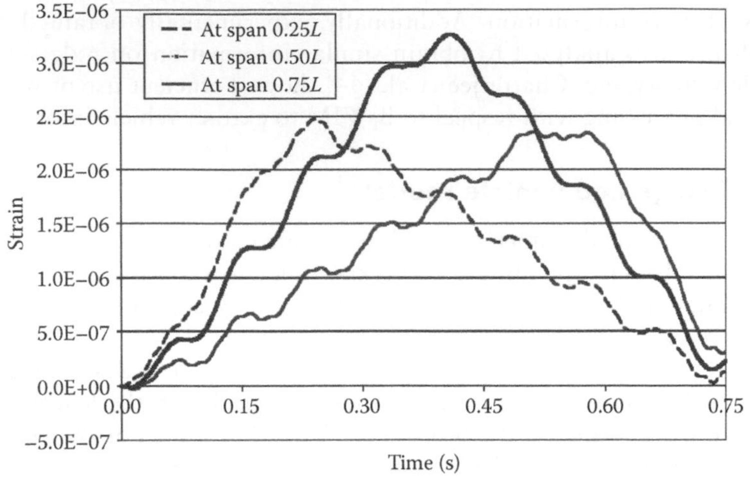

Figure 7.2 Strain signals at spans 0.25*L*, 0.50*L* and 0.75*L*.

or extract any other information from the generated strain signals. The original strain signals at 0.25*L*, 0.50*L* and 0.75*L* on the beam model are shown in Figure 7.2. The wavelet analysis of the strain signals is done in MATLAB®. A reverse biorthogonal wavelet basis function is chosen here to analyse the strain signals. After performing a number of analyses using other existing wavelet basis functions in MATLAB® (which are shown later), a reverse biorthogonal wavelet *rbio*2.4 is selected to analyse the strain signals and also the other NOR signals later in this chapter. The wavelet *rbio*2.4 is a compactly supported biorthogonal spline wavelet with symmetry and perfect reconstruction properties. Unlike the case of the orthogonal wavelet, where the analysis wavelet function is the same as the synthesis wavelet function (hence the number of vanishing moments is a single value), two sets of low-pass and high-pass filters are used in the biorthogonal wavelet transform – one is used for decomposition and the other for reconstruction (i.e. analysis and synthesis, respectively).

The numbers 2 and 4 in *rbio*2.4 denote the orders (vanishing moments) for decomposition and reconstruction of the associated wavelet function, respectively. The biorthogonal wavelets are symmetric and have compact support. The decomposition and reconstruction wavelet functions (*rbio*2.4) are shown in Figures 7.3 and 7.4. The biorthogonal wavelets retain more energy than the orthogonal wavelets in lower-frequency subbands.

Using this wavelet function, *rbio*2.4, the strain signal, is first decomposed and reconstructed using a standard MATLAB® function, and the contour of wavelet coefficients is obtained at different scales, as shown in Figure 7.5.

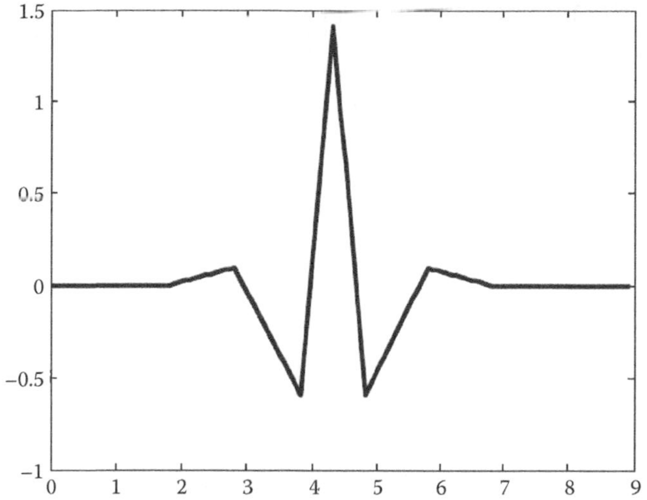

Figure 7.3 Decomposition function for mother wavelet *rbio*2.4.

The two prominent vertical lines at about 0.18 and 0.25 s indicate the axle positions, and these are much pronounced at lower scales (i.e. higher frequencies). The wavelet coefficients at a particular scale are shown in the time domain in Figure 7.6, in which the passage of each axle can be easily identified from two sharp peaks. The time difference between the occurrence of

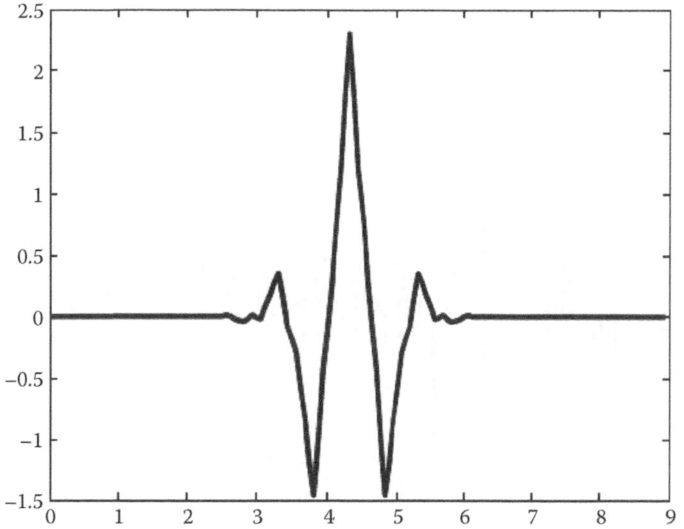

Figure 7.4 Reconstruction function for mother wavelet *rbio*2.4.

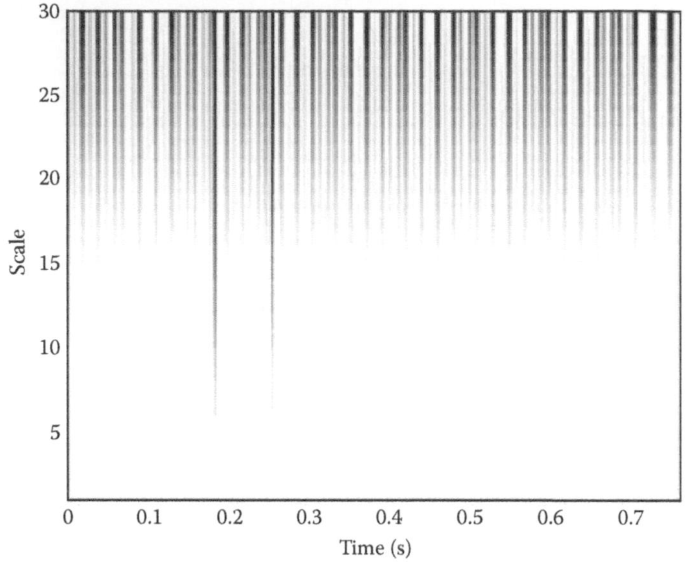

Figure 7.5 Wavelet coefficients at 0.25*L* for all scales considered.

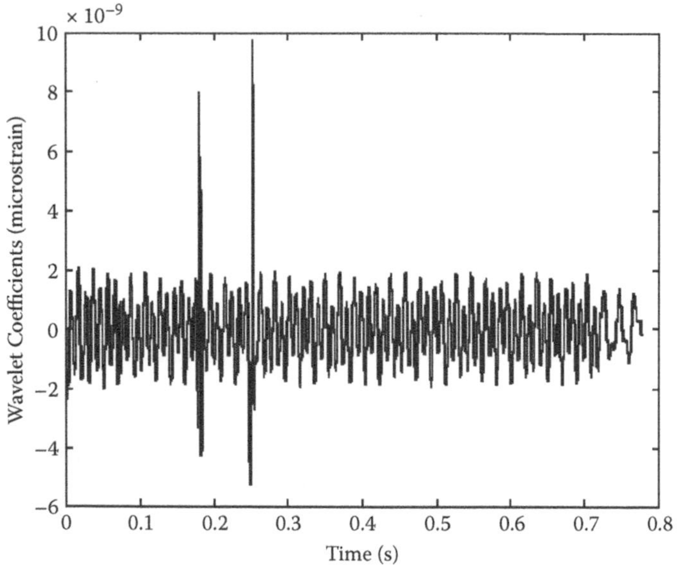

Figure 7.6 Wavelet coefficients at 0.25*L* for a specific scale (scale 16).

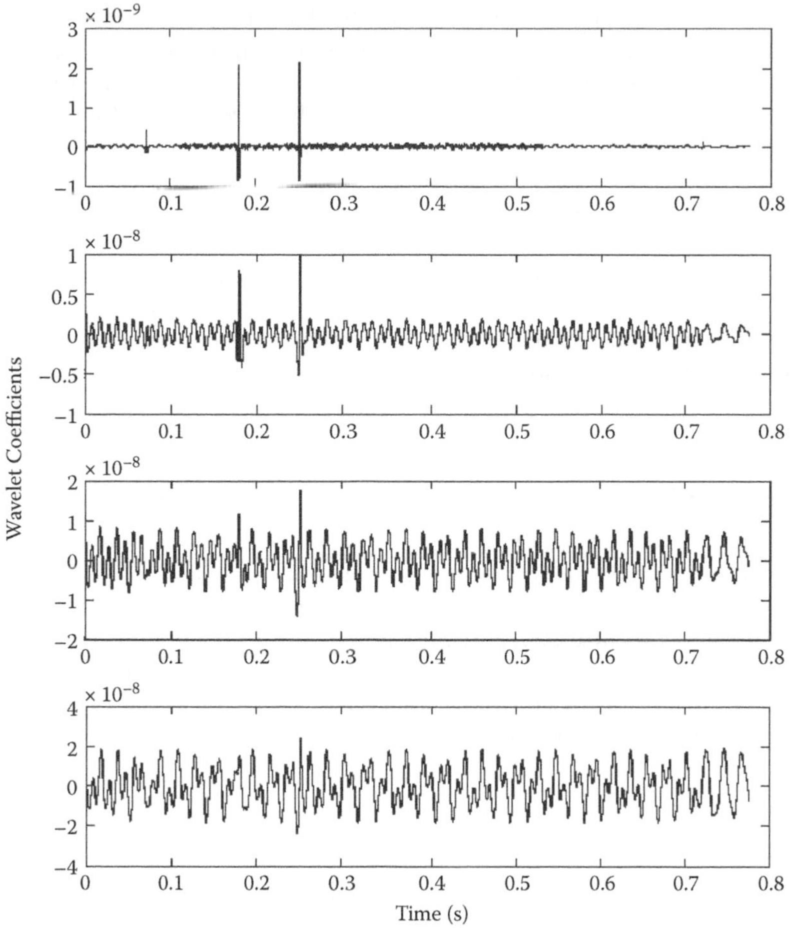

Figure 7.7 Wavelet coefficients at 0.25*L* for different scales.

these two peaks accounts for the difference in time of arrivals of the first and second axles at 0.25*L* and 0.75*L*. The corresponding velocity may be subsequently calculated to be 74.85 km/h, which is only 0.2% different from the actual value of 75 km/h. From this velocity and the distance between the peaks corresponding to each axle, the axle spacing comes out to be 1.5012 m, indicating an error of only 1.2 mm from the 1.5 m spacing used.

Figure 7.7 demonstrates the wavelet coefficients at 0.25*L* for four different scales (6, 16, 24 and 32). This figure reveals that at higher scales the peaks are getting lost. The wavelet used here is effective at isolating the axle passages from the effects of vehicle and bridge vibrations. Thus, it is seen that the details of the vehicle passage can be identified from wavelet-based analysis of the original strain signal, which would otherwise be quite difficult. In this case, the original strain signal has been derived numerically from a physical model with known characteristics. Thus, the foregoing example may be considered a validation exercise to prove the suitability of wavelet analysis to process a nonstationary random signal due to vibration induced by the passage of a vehicle. This gives confidence to analyse similar signals obtained from experiments conducted at sites. Now, we will try to perform wavelet-based analysis of some real-time measurements of vibrations of bridges (in the form of strain signals) due to axle passages and extract similar information.

7.1.2 Wavelet analysis of experimental NOR data

Chatterjee et al. [44] has referred to a series of experimentally obtained NOR strain signals measured at different sensors placed underneath the Ravbarkomanda box culvert in Slovenia in 2004. A total of 16 channels were used to measure NOR strain signals at a frequency of 512 Hz. Of these channels, six were placed at mid-span to determine axle weights and two were placed in the slow lane under the wheel track and off-centre to correctly identify the axles of vehicles passing by. Figure 7.8 shows the

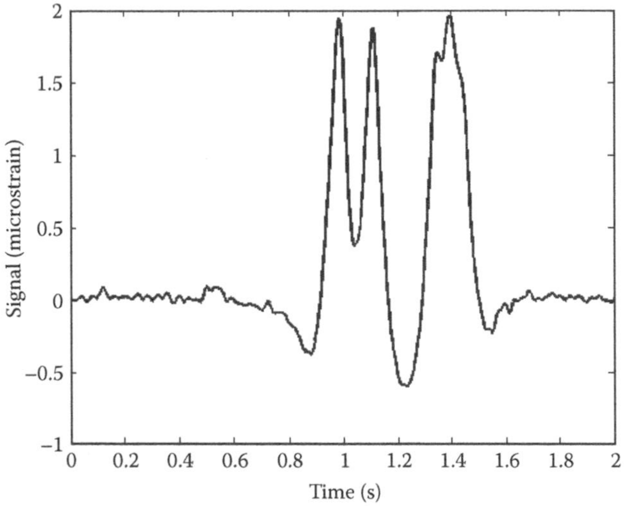

Figure 7.8 Original strain signal.

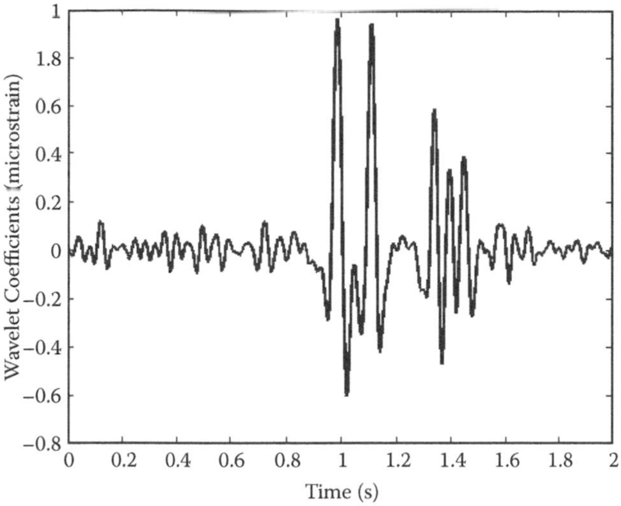

Figure 7.9 Wavelet-transformed (using *rbio*2.4) original signal at scale 16.

original strain signal measured at one of those two sensors placed in the slow lane corresponding to the passage of a truck with a gross vehicle weight of about 179 kN with five axles. The vehicle was stopped to measure the axle distances with tape, and these are as follows. The distances between the first and second axles and the second and third axles are 3.19 and 6.06 m, respectively. The third to fourth axle distance and the fourth to fifth axle distance are measured to be 1.32 and 1.31 m, respectively.

The original strain signal shows peaks corresponding to four axles with a very weak third peak. However, the fifth axle is hardly seen; rather, a negligible change in the slope of the signal is visible. This signal is analysed using continuous wavelet transform in MATLAB®, and the wavelet basis function chosen is *rbio*2.4. The wavelet coefficients are computed at various scales. The wavelet-transformed strain signal corresponding to scale 16 is shown in Figure 7.9. All five axles can now be identified from the spectrum very clearly. This is due to the unique time–frequency localization property of the wavelets that the time information is not lost over different scales or frequencies, unlike Fourier transform. The wavelet transformation is able to retain the original relationship with time. However, the level of prominence of peaks may depend on the choice of scale. In this case, scale 16 is seen to be the most useful one, and at the same time, it has been verified that the wavelet coefficients corresponding to other scales cannot produce such sharp peaks indicating axle positions so distinctly. The ripples in the transformed signal may be attributed to the fact that the changes in

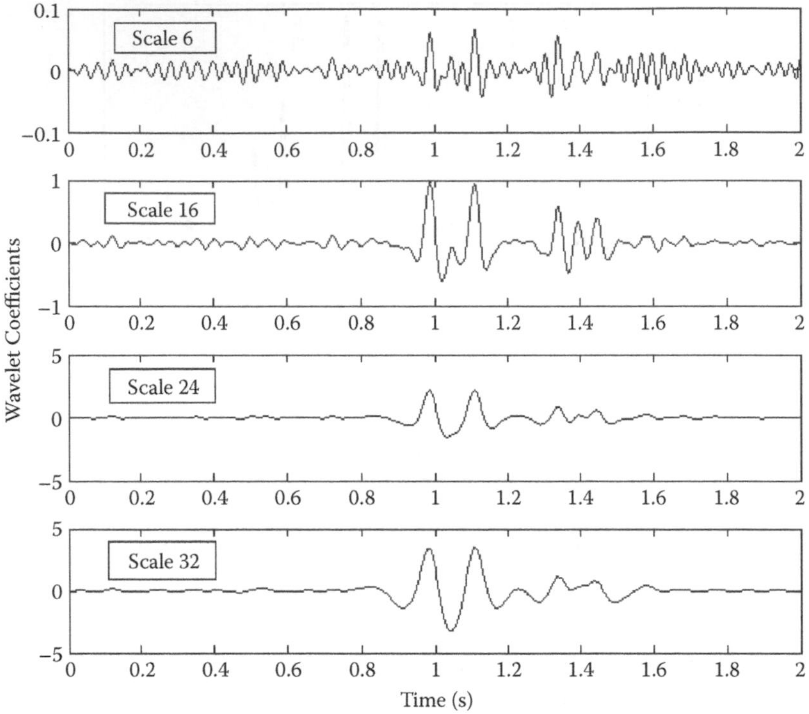

Figure 7.10 Wavelet-transformed (using *rbio*2.4) original signal at various scales.

frequencies might be sudden. The same data, when analysed with other wavelets, such as the Daubechies family of wavelets, did not yield results as good as shown in Figure 7.9. To give readers a reasonable comparison, the wavelet-transformed signal of the original strain signal using *rbio*2.4, but at different scales, is shown in Figure 7.10. It is clearly visible from this figure that the wavelet coefficients at scale 16 are able to show all the peaks representing all vehicle axles correctly in the time domain and wavelet coefficients at other scales cannot reveal this information. The axle spacing may be subsequently calculated from the position of the peaks in the time domain.

The same, using the Daubechies family of wavelets *db*6, *db*12 and *db*18, is shown in Figures 7.11 to 7.13, respectively. The figures demonstrate that the Daubechies family of wavelets is not able to identify axles distinctly at any scale. So, the selection of the correct wavelet function for the analysis is very important.

The wavelet-based analysis of the signal thus helps in identifying the arrival of vehicle axles in the time domain, correct axle spacing and hence the velocity of the vehicle.

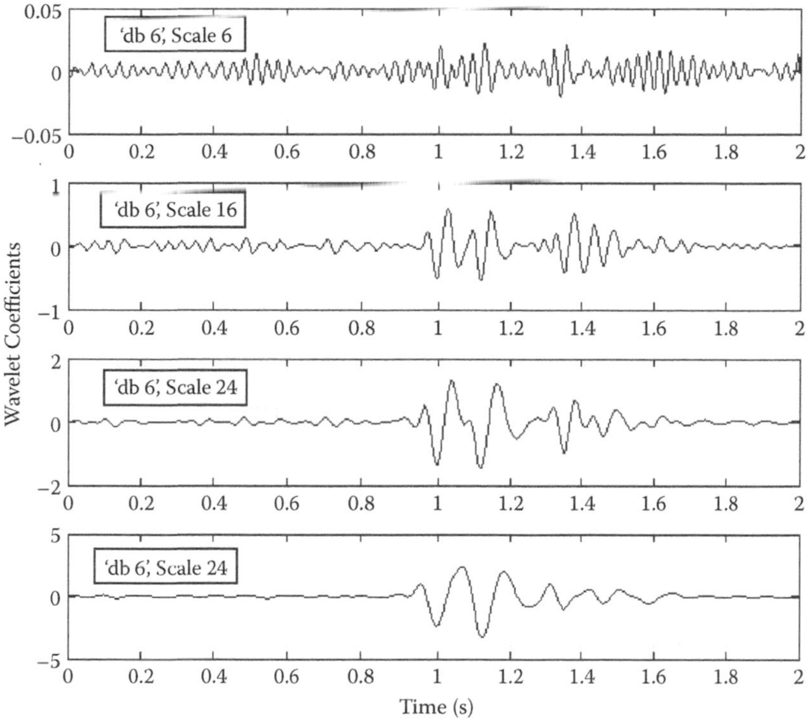

Figure 7.11 Wavelet-transformed (using *db*6) original signal at various scales.

7.2 STIFFNESS DEGRADATION ANALYSIS

A reliable and popular approach toward structural health monitoring and damage detection is based on the wavelet analysis of structural response data. It has been shown in numerous studies that the effect of system degradation is reflected in time–frequency characteristics of the response. The wavelet coefficients of the response of the structure reveal important information on stiffness degradation and help sort out the damage locations. The cracks developed in engineering structures may cause local variation in stiffness, and hence affect the nature and type of the response of the structure. The cracks affect the natural frequencies and mode shapes of the structure. The powerful wavelet analytic tool helps in damage detection of affected structure and gives precise information about the temporal variation of the frequency content of the nonstationary response of the cracked structure. A simple example of a mass-spring-dashpot system representing a structural frame subjected to earthquake base excitation is analysed, and

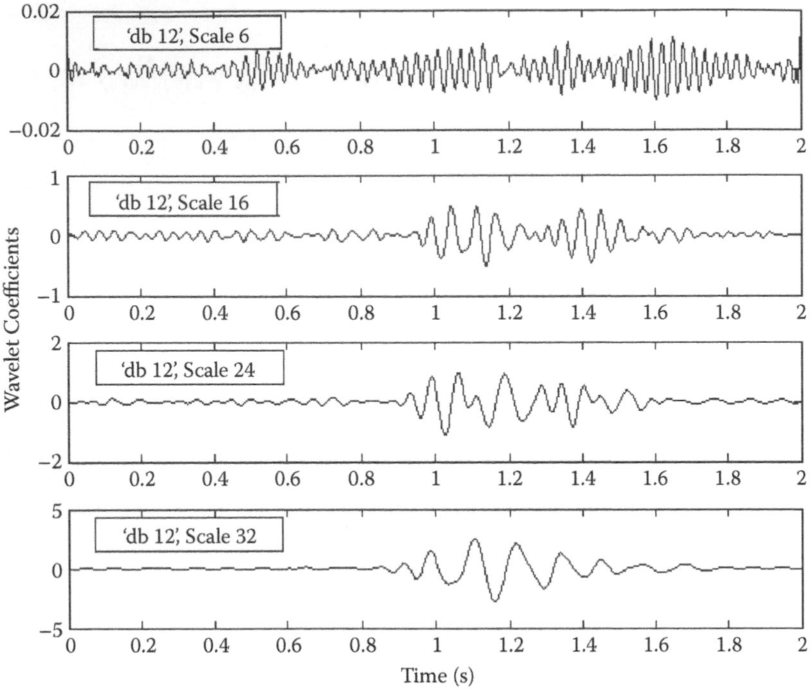

Figure 7.12 Wavelet-transformed (using *db*12) original signal at various scales.

the response is obtained, which is nonstationary in nature. The wavelet-based analysis of the response is carried out to show the prominent effects of stiffness degradation.

7.2.1 Description of the analytical model

Let us first consider a simple model of a structure represented by the combination of a single mass connected to multiple springs in parallel and a single viscous dashpot. The springs are analogous to the stiffness of the structural components (beams, columns, etc.). The model is shown in Figure 7.14. The displacement of the mass is denoted by $x(t)$, and the mass is subjected to an external time-varying force $P(t)$. The springs have stiffness coefficients denoted by K_1, K_2, K_3, etc., and the damper is assumed to be damping coefficient C. Each spring is assumed to exhibit a nonhysteretic bilinear behaviour. Hence, each spring is given a specific value of displacement (a threshold) that when exceeded, it breaks. Once the spring fails, it does not contribute to the system stiffness any more. However, if the force in the spring comes below the threshold value, it offers resistance again. The

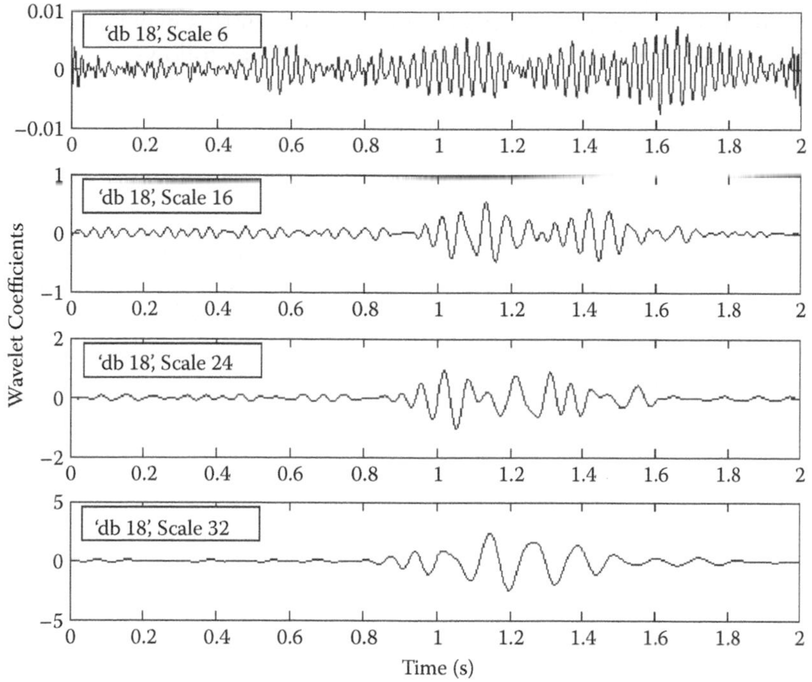

Figure 7.13 Wavelet-transformed (using *db*18) original signal at various scales.

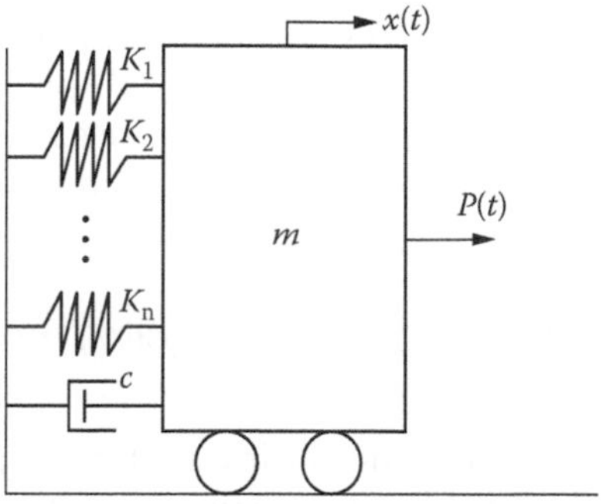

Figure 7.14 Multilinear-spring–dashpot-supported structural mass.

failure of a spring may be compared to the damage of a structure. The system stiffness is time varying, and at any time instant this may be expressed as the summation of individual spring stiffness, K_j, as follows:

$$K(t) = \sum_{j=1}^{n} K_j(t) \tag{7.3}$$

This actually represents a multilinear nonhysteretic concrete structure. The time-varying stiffness of any individual spring, $K_j(t)$, follows a bilinear path, depending upon a predefined yield displacement, x_y, as follows:

$$K_j(t) = \begin{matrix} K_i\,, & \text{if } |x(t)| < x_y \\ 0, & \text{otherwise} \end{matrix} \tag{7.4}$$

The term K_i denotes the initial spring stiffness.

The equation of motion for this mass–multilinear-spring–single-dashpot combination is written below:

$$m\ddot{x}(t) + 2m\zeta\omega_n(t)\dot{x}(t) + m\omega_n^2(t)x(t) = m\ddot{x}_i(t) \tag{7.5}$$

In this dynamic equation, the term $\ddot{x}_i(t)$ denotes the input acceleration, $\omega_n(t)$ is the natural frequency of the system and varies with time and ζ is the damping ratio, which is also time dependent. Equation (7.5) may be solved in the time domain to obtain the response of the system. Once the solution is obtained, the response may be inspected in more detail using wavelet coefficients, and the results may then be compared with the wavelet coefficients of the input excitation.

In order to obtain a more realistic representation of a structural system that behaves nonlinearly at low vibration amplitude, unlike the model discussed above, a continuous hysteretic force displacement curve may be assumed. This allows the stiffness to vary at every time instant (unlike bilinear variation in the above case). The equation of motion in this case is described as [45]

$$m\ddot{x}(t) + 2m\zeta\omega_n(t)\dot{x}(t) + m\chi\omega_n^2(t)x(t) + (1-\chi)\omega_n^2(t)z(t) = m\ddot{x}_i(t) \tag{7.6}$$

In this equation, one can see some additional terms appearing. The terms $z(t)$, $x(t)$, ζ, ω_n and $\ddot{x}_i(t)$ denote, respectively, the hysteretic displacement, structural displacement, linear damping ratio, natural frequency of the system and input acceleration. The term χ represents the rigidity ratio of the hysteretic force-displacement loop. For a linear system, the value of χ is

1. For a highly nonlinear system, this value should be relatively smaller, and for $\chi = 0$, the system is completely nonlinear. The third term, $m\mathcal{X}\omega_n^2(t)x(t)$, and the fourth term, $(1-\mathcal{X})\omega_n^2(t)z(t)$, are linear restoring force and hysteretic force, respectively. The relation between the structural displacement and the hysteretic displacement is given below:

$$\dot{z}(t) = -\gamma|\dot{x}(t)|z(t)|z(t)|^{m-1} - \lambda\dot{x}(t)|z(t)|^{m} \qquad (7.7)$$

The basic shape of the hysteresis is determined by three shape parameters, γ, λ and m, as can be seen in Equation (7.7).

7.2.2 Numerical approach to wavelet-based damage detection

Unlike in previous chapters, this time we will not solve Equations (7.6) and (7.7) in the wavelet domain. The guideline on formulation of basic equations of motion in case of damage analysis is given in the previous section. These equations can be solved numerically. However, when the model is big and quite complicated, and especially for three-dimensional problems, a finite element model of the structural model is likely to be developed. The response of the structural system may be subsequently obtained once the material properties, the boundary and support conditions and the loading conditions are appropriately defined and assigned to the right parts of the finite element meshes. Once the responses are obtained, these can be analysed further computing wavelet coefficients. One simple example as given below may illustrate the efficiency of the wavelet-based signal processing technique in revealing some important information regarding the formation of cracks in the structure or stiffness degradation of the structure, which would have been more difficult to ascertain from ordinary Fourier transform of the data or analysis by any other technique.

7.2.3 Finite element model

A two-dimensional numerical model of a shear wall fixed at the bottom is developed using finite elements (software DIANA, version 9.5). The model of the wall is shown in Figure 7.15. The wall panel is 0.1 m thick. It has vertical flanges on both sides and supports a beam at the top. The reinforcements are assumed to be present in the wall and the flange (not in the beam). The base of the model is subjected to a random seismic ground motion process simulating the Loma Prieta ground motion as shown in Figure 3.1. The base excitation is applied in both the directions (vertical and horizontal)

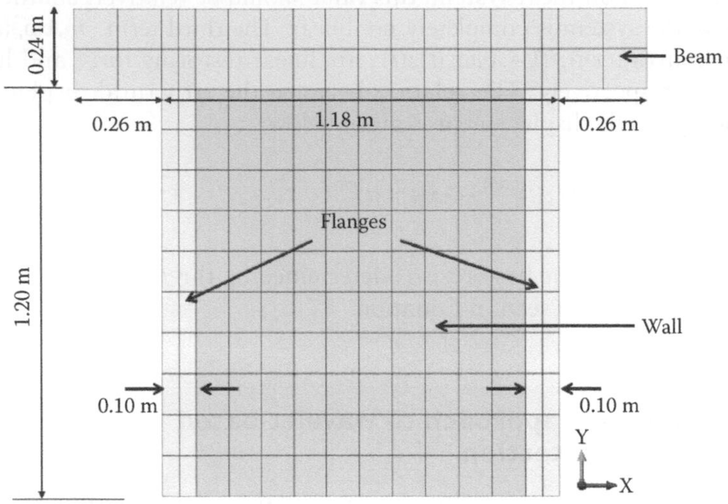

Figure 7.15 Model of shear wall panel.

in the model simultaneously. The model dimensions are clearly shown in Figure 7.15. The mesh in this model is developed using eight-node quadrilateral isometric plane stress elements. The whole structure is made up of concrete with reinforcements of steel wherever applied. A number of runs have been made assuming linear and nonlinear material properties for the concrete in the same model as described in Table 7.1. The steel reinforcements were always assumed to follow linear behaviour. In order to simulate development of cracks, nonlinear material properties were assumed to follow a total strain rotating (orthogonal) crack model, which is usually applied to the constitutive modelling of reinforced concrete during a long period and quite well suited to reinforced concrete structures. A five-point multilinear tension curve, represented in Figure 7.16, is used to define fully the stress-strain relationship, which has made the input of the tensile strength redundant, and this is a more pragmatic approach for a numerical model. The crushing behaviour of the concrete is simulated by a constant compression curve, as shown in Figure 7.17.

Table 7.1 Material properties for concrete and steel

	Unit	*Concrete*	*Steel*
Young's modulus	N/m^2	3×10^6	2×10^{11}
Poisson's ratio	—	0.15	0.2
Density	kg/m^3	2400	7850

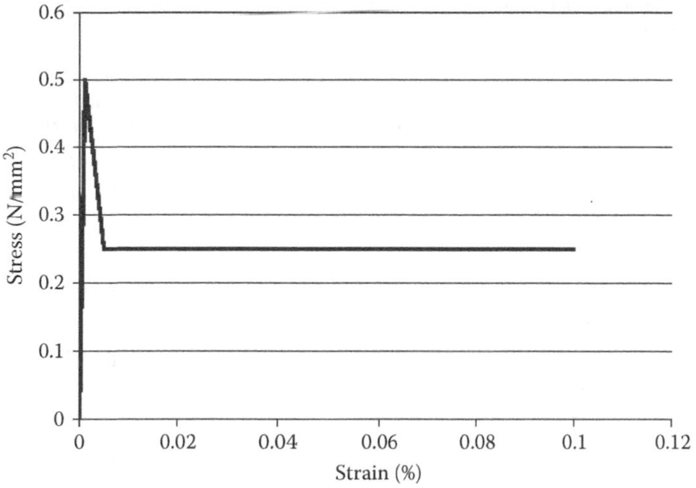

Figure 7.16 Multilinear tension curve for the wall.

To make the model more realistic, Rayleigh damping parameters are introduced into the model. Actually, in dynamic analysis of structures and foundations, damping significantly affects the response. The most effective and relatively easier way to tackle damping is to consider it within the modal framework. To obtain the value of the Rayleigh damping parameters (α and β) that depend on a pair of dominant eigenfrequencies (ω_i) and corresponding damping ratios (ζ_i), as shown by the Equation (7.8), a free vibration analysis (also referred to as eigenanalysis) of the model is

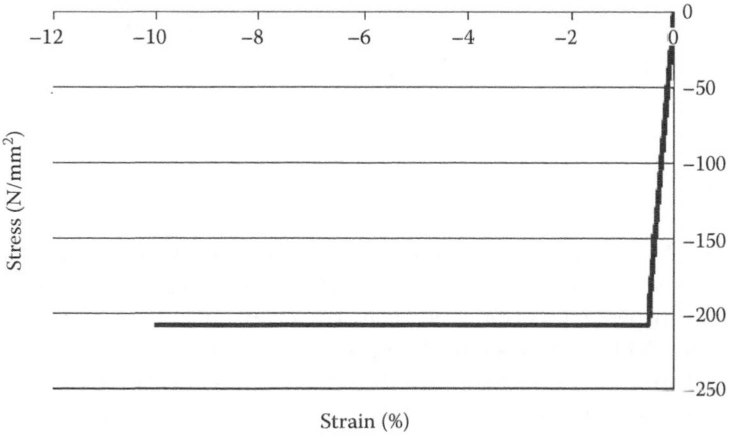

Figure 7.17 Compression curve for the crushing concrete.

Table 7.2 Eigen frequencies and mass participation factors obtained from modal analysis

Mode	*Frequency (Hz)*	*Cumulative mass participation (%) – horizontal direction*	*Cumulative mass participation (%) – vertical direction*
1	11.54	0.00	83.46
2	12.07	1.78	83.46
3	17.75	1.78	86.93
4	21.39	84.13	86.93
5	28.62	87.27	86.93
6	29.98	89.10	86.93
7	33.30	89.10	86.98
8	39.27	89.10	87.59
9	45.33	89.10	89.26
10	47.36	92.08	89.26
11	50.52	92.08	89.98
12	54.11	92.08	94.47
13	54.62	92.11	94.47
14	66.19	92.34	94.47
15	69.85	93.47	94.47

first carried out. This analysis yields some important information listed in Table 7.2.

$$2\zeta_i\omega_i = \alpha + \beta\omega_i^2; \quad i = 1,2 \tag{7.8}$$

The terms α and β are used subsequently in the numerical model to compute the damping matrix [C] of the physical system with the help of the following expression, in which [M] denotes the mass matrix of the system and [K] represents the system stiffness matrix.

$$[C] = \alpha[M] + \beta[K] \tag{7.9}$$

Table 7.2 shows the natural frequencies of free vibration of the shear wall in both directions (horizontal, X; vertical, Y) and the cumulative mass participation in percent as obtained from the eigenvalue analysis of the finite element model for this system, shown in Figure 7.15.

From Table 7.2 it is clearly seen that modes 4, 6 and 10 are dominant in the X direction and modes 1, 3 and 12 are dominant in the Y direction. Hence, from corresponding eigenfrequencies it is observed that the dominant frequency range lies between 11 and 55 Hz. For assumed damping

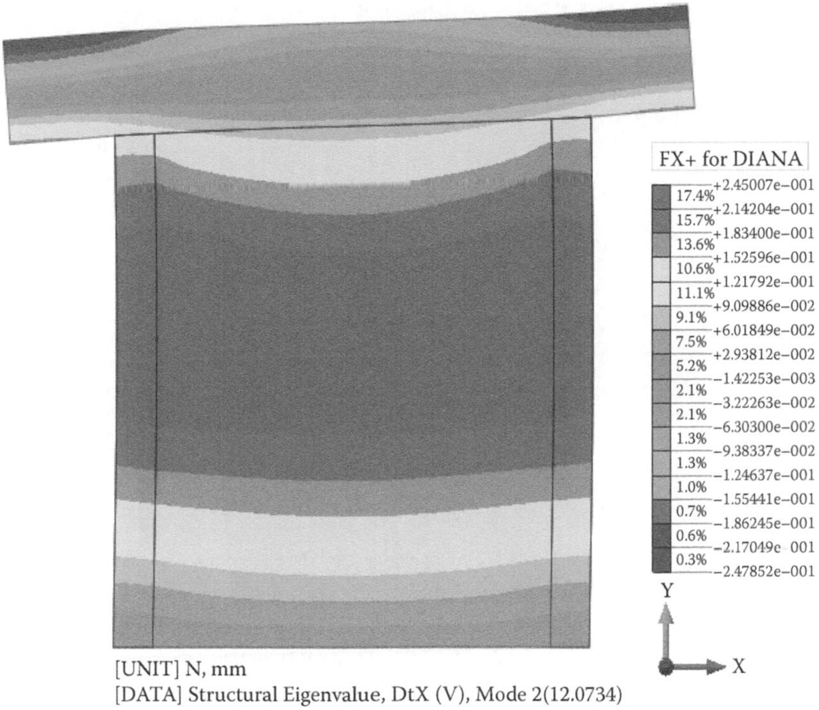

Figure 7.18 Mode shape 2 at 12.07 Hz.

of 2%, the Rayleigh damping parameters, α and β, have been evaluated to be 5.762 and 0.00024, respectively, from Equation (7.8). These values are used as input parameters to the model when it is necessary to introduce 2% damping. Two sets of results for the linear model have been obtained – one for the 'no damping' condition and the other with 2% damping. Two sets of results were further obtained in the case of the nonlinear concrete model (using nonlinear material properties for cracks as discussed above) for the no damping condition and with 2% damping. Some of the mode shapes corresponding to dominant frequencies in the lateral (X) direction obtained from the modal analysis of the finite element model are shown in Figures 7.18, 7.19, 7.20 and 7.21. The results were obtained in terms of acceleration response of the system and are illustrated in the next section.

7.2.4 Wavelet-based analysis of numerical results

The responses were obtained from the numerical model in terms of the acceleration response of specific nodes in the system. These nodes are

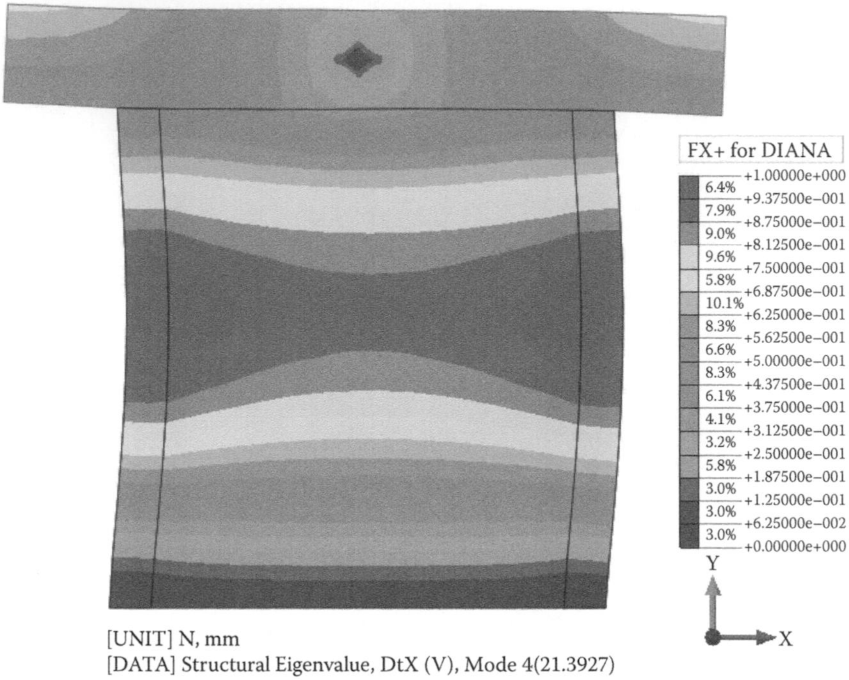

Figure 7.19 Mode shape 4 at 21.39 Hz.

chosen from the area that suffered maximum cracks in the case of the nonlinear model. Figures 7.22 and 7.23 show the temporal variation of the response of the current model in terms of acceleration of the linear and nonlinear models, respectively, obtained at a particular node (precisely node number 102) that lies within the most affected area suffering the maximum number of cracks. Apparently, these two acceleration time histories do not reveal any good information about the system. The wavelet coefficients are computed for the input excitation (Loma Prieta) and also for the response for linear and nonlinear systems (both with 2% damping) and are plotted in Figures 7.24 to 7.26, respectively. The wavelet coefficients of acceleration response for the linear model clearly seem to be different from the wavelet coefficients of the acceleration response of the nonlinear model. The amplitudes of wavelet coefficients of the acceleration response for the nonlinear model seem to be higher than those of the linear model. The cross-correlation coefficients are computed between input excitation and linear model and also between input excitation and nonlinear model and are demonstrated, respectively, in Figures 7.27 and 7.28. It is vividly seen that the coefficients are closer to zero in the case where the nonlinear

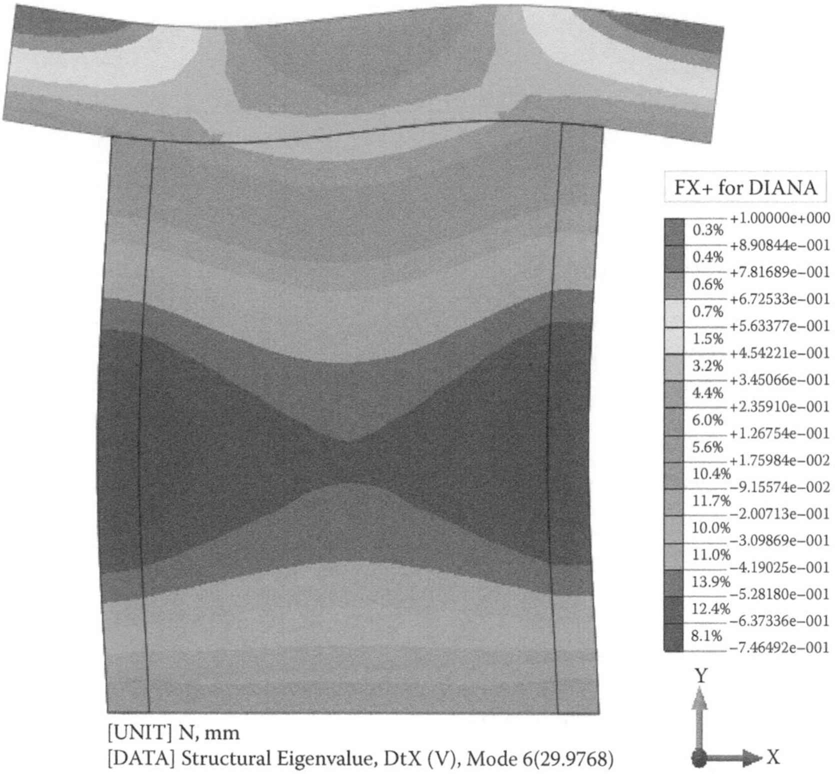

Figure 7.20 Mode shape 6 at 29.98 Hz.

model is considered, which indicates that stiffness degradation is higher in the case of the nonlinear model than in the case of the linear model.

7.3 SOIL–STRUCTURE–SOIL INTERACTION ANALYSIS

In the previous chapters, we have seen methods considering nonstationarities in ground motions in wavelet-based analysis for soil–structure interaction. We have cited a few examples on soil-building interaction and soil-tank interaction, including some numerical illustrations. Fluid–soil–structure interaction analysis and soil–structure interaction analysis due to nonstationary random excitations have been carried out by numerous researchers in the past, as referred to in previous chapters. In fact, some interesting works have also been presented recently by Livaoglu [46], Tam [47], Ding and Sun [48], Ovanesova and Suárez [49] and others. The advantage of wavelet-based analysis lies in the fact that it may efficiently account

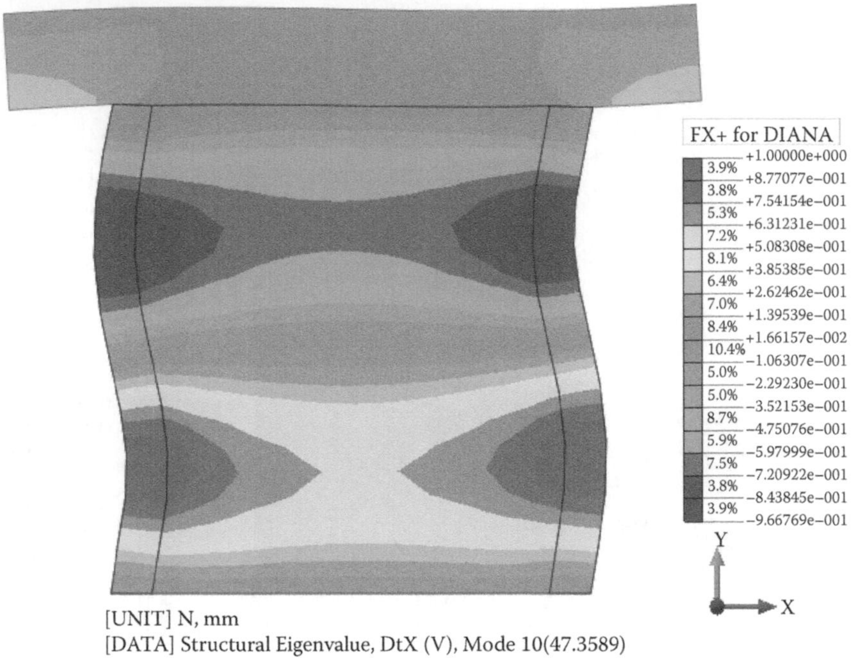

Figure 7.21 Mode shape 10 at 47.36 Hz.

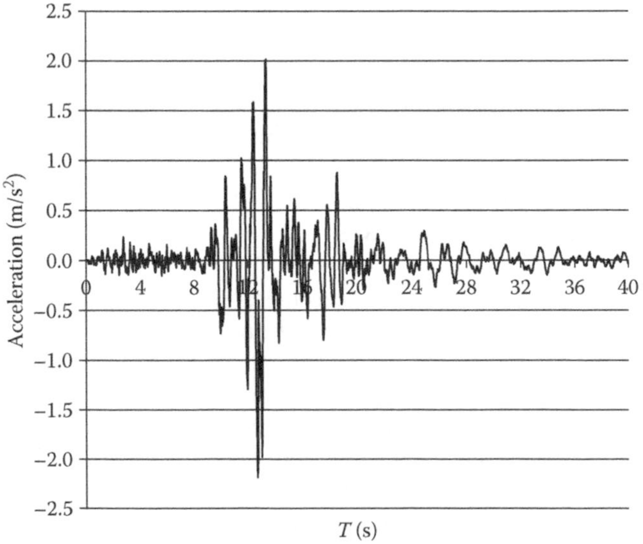

Figure 7.22 Acceleration response of linear model with 2% damping.

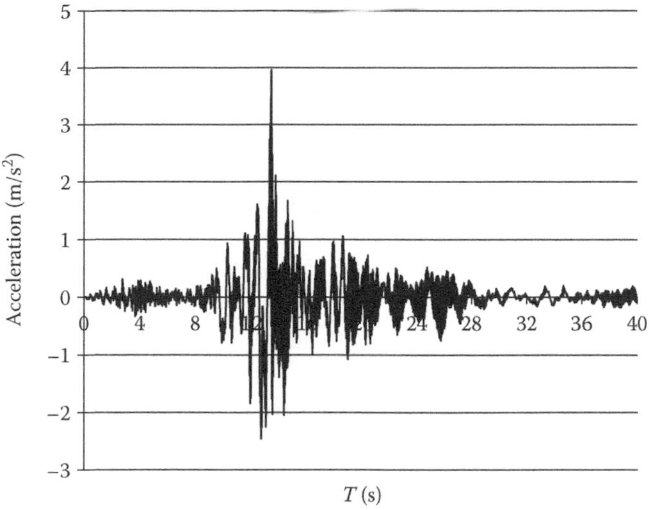

Figure 7.23 Acceleration response of nonlinear model with 2% damping.

for the nonstationarities in input excitation. In the formulation discussed in Chapters 2 to 6, the responses of the structural models were obtained through wavelet coefficients. In most of the cases, the analysts are interested in obtaining only the response of the structure due to ground shaking. However, at certain times, the structural vibrations may have a nonnegligible effect on the response of the underlying soil medium. In this section, a simple technique to obtain a soil response considering nonstationary soil–structure–soil interaction is presented. For the wavelet domain analysis of

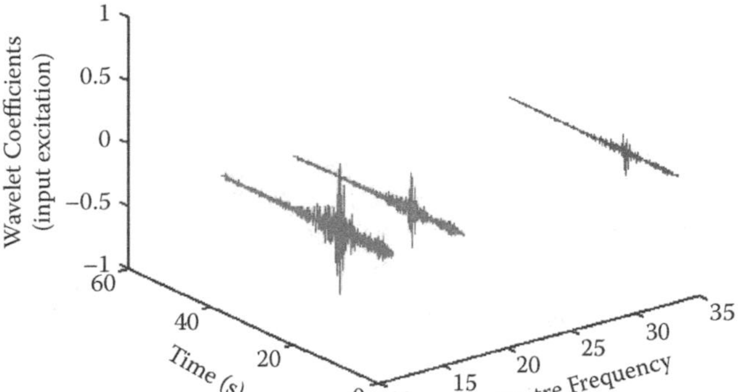

Figure 7.24 Wavelet coefficients of input seismic acceleration (Loma Prieta).

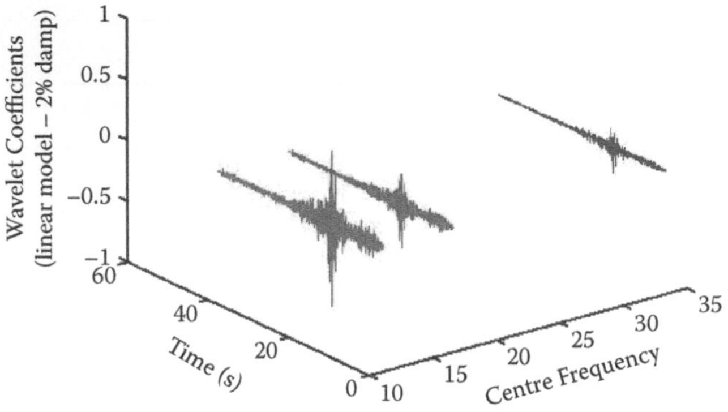

Figure 7.25 Wavelet coefficients of acceleration response of node 102 in linear model.

the physical model (single degree of freedom (SDOF)) of the system, as shown in Figure 7.29, the technique remains the same as used in the previous chapters. This time, we solve the system dynamics three times to obtain three pairs of shear and moment at the base for the same tank for three different accelerograms. A tank has been considered to be the superstructure and is imagined to be placed at three different locations having three different types of earthquake record. The three different input ground accelerations are representatives of seismic ground motions at Loma Prieta (United States, 1989), Miyagi (Japan, 1978) and El Centro (United States, 1940). These accelerograms are shown in Figures 7.30, 7.31 and 1.2. A 3D finite element model of the soil domain with a centrally placed circular footing

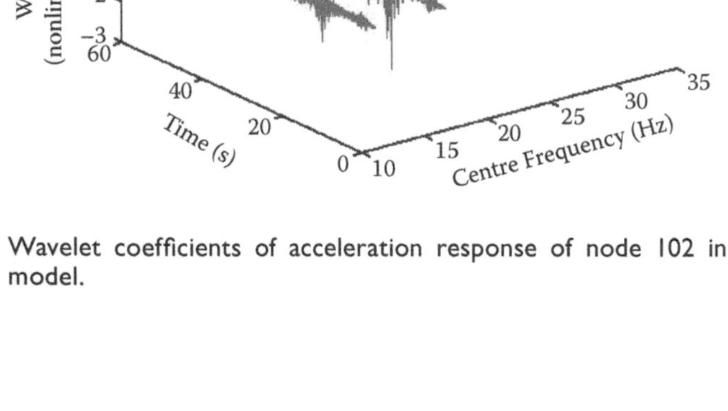

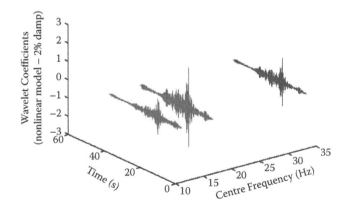

Figure 7.26 Wavelet coefficients of acceleration response of node 102 in nonlinear model.

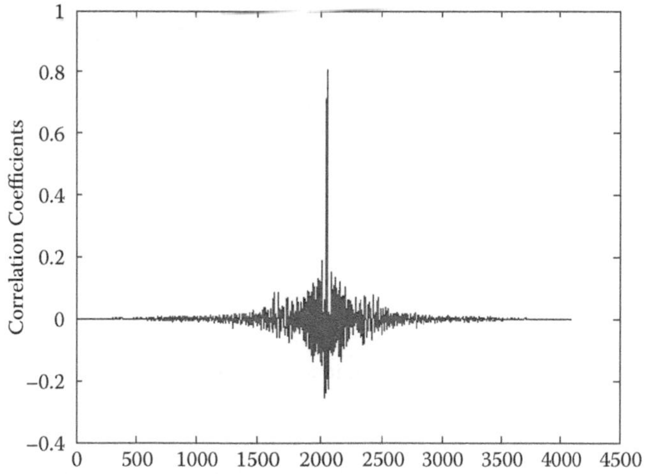

Figure 7.27 Cross-correlation coefficients between input excitation and linear model response.

on the top of the surface is subsequently developed in DIANA. The values of base shear and base moment as obtained from wavelet-based analysis are applied simultaneously to the nodes on the surface of the footing in the appropriate direction. The 3D nonlinear static analysis of the finite element model is then carried out to obtain the values of displacements, strains and

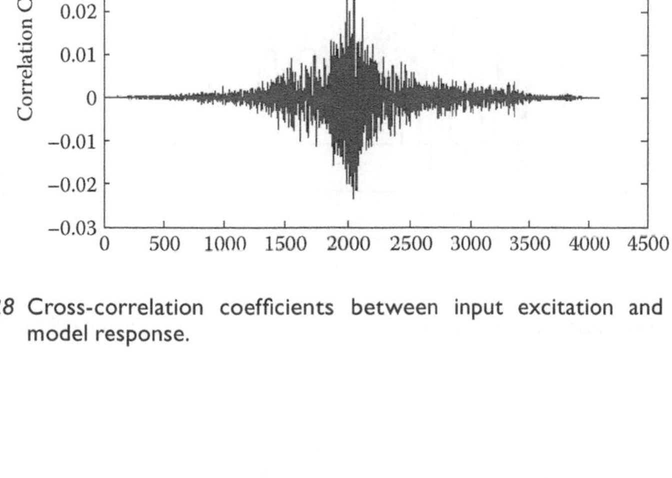

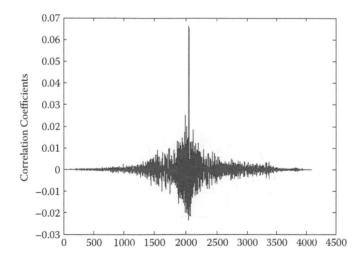

Figure 7.28 Cross-correlation coefficients between input excitation and nonlinear model response.

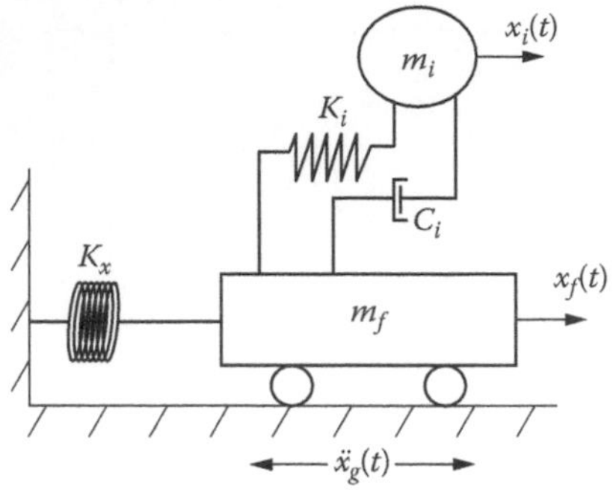

Figure 7.29 2-DOF model of the tank–foundation system.

stresses. The whole process is repeated three times – each time with a different accelerogram.

7.3.1 Responses at tank base

We consider here a simple model of a tank–foundation–soil structure. The model is shown in Figure 7.29. The physical characteristics of the system are represented by impulsive mass (m_i), foundation mass (m_f), spring stiffness K_i connecting the liquid to the foundation, damper C_i and complex-valued impedance of the soil-foundation system (K_x). The displacements of the impulsive mode of the tank–liquid and the foundation are denoted

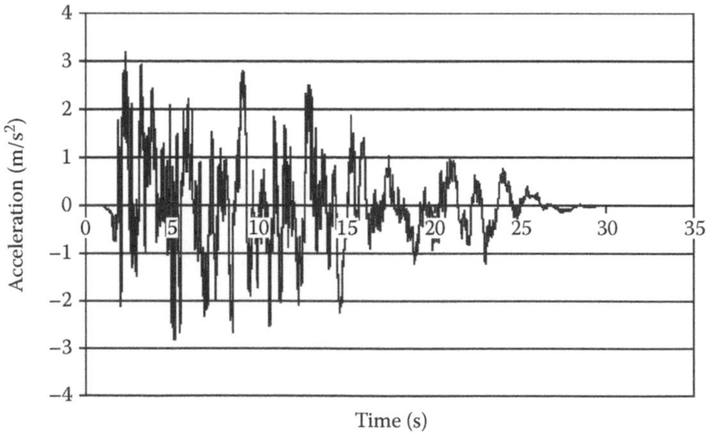

Figure 7.30 Time history of acceleration during the Miyagi earthquake (1978), Japan.

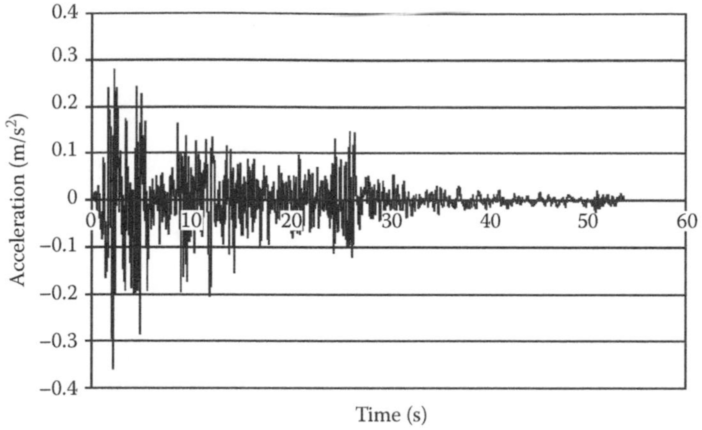

Figure 7.31 Time history of acceleration during the El Centro earthquake (1940), United States.

by $x_i(t)$ and $x_f(t)$, and these are relative to the ground displacement. The whole system is subjected to ground shaking with acceleration time history denoted by $\ddot{x}_g(t)$. The equations of motion may be formulated for linear soil–structure interaction analysis and subsequently solved using the wavelet-based technique as explained earlier. For the sake of the readers' interest, the dynamic equations of the tank–fluid–foundation system and only the tank–fluid system may be written as

$$m_i(\ddot{x}_i(t)+\ddot{x}_f(t))+m_f\,\ddot{x}_f(t)+K_x(x_f(t)-x_g(t))=0 \tag{7.10}$$

$$m_i\ddot{x}_i(t)+C_i\dot{x}_l(t)+K_i\,x_i(t)=-m_i\ddot{x}_i(t) \tag{7.11}$$

It may be noted here that in the above equation, $C_i = 2\zeta m_i\omega_i$ and $K_i = m_i\omega_i^2$, and the expression for the term K_x is given by Equation (4.16). The expressions for the base shear, $Q(t)$, and the base moment, $M(t)$, developed at the tank base due to soil–structure interaction effects may be written as

$$Q(t)=m_i\left\{\ddot{x}_i(t)+\ddot{x}_f(t)\right\} \tag{7.12}$$

$$M(t)=m_i\left\{\ddot{x}_i(t)+\ddot{x}_f(t)\right\}h_i \tag{7.13}$$

In the above expression for tank base moment the term h_i denotes the height above the tank base at which the impulsive mass should be located to give correct moments at a section immediately above the tank base. The

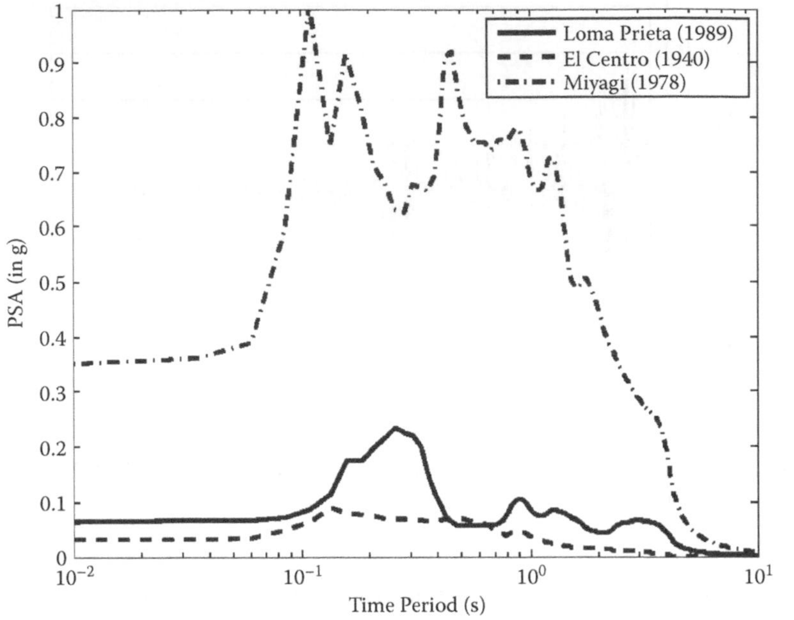

Figure 7.32 PSA response of a SDOF system for three different accelerograms.

base shear and the base moment may also be expressed in terms of wavelet coefficients as

$$Q(t) = \sum_i \sum_j \frac{K''\Delta b}{a_j} W_\psi Q(a_j, b_i)\psi_{a_j,b_i}(t) \tag{7.14}$$

$$M(t) = \sum_i \sum_j \frac{K''\Delta b}{a_j} W_\psi M(a_j, b_i)\psi_{a_j,b_i}(t) \tag{7.15}$$

The reader may now follow the procedure adopted in previous chapters in solving the system equations in the wavelet domain. Without getting into the details of the solution method (which is already cited in previous chapters), the expressions for impulsive displacement and foundation displacement responses in terms of functionals of wavelet coefficients are given below:

$$\sum_i W_\psi x_i(a_j, b_i)\hat{\psi}_{(a_j,b_i)}(\omega) = \tau_i(\omega)\sum_i W_\psi \ddot{x}(a_j, b_i)\hat{\psi}_{(a_j,b_i)}(\omega) \tag{7.16}$$

$$\sum_i W_\psi x_f(a_j, b_i)\hat{\psi}_{(a_j,b_i)}(\omega) = \tau_f(\omega)\sum_i W_\psi \ddot{x}_g(a_j, b_i)\hat{\psi}_{(a_j,b_i)}(\omega) \tag{7.17}$$

In the above equations, (7.16) and (7.17), $\tau_i(\omega)$ and $\tau_f(\omega)$ are the frequency-dependent complex-valued transfer functions relating the wavelet coefficients of horizontal displacements of impulsive tank–liquid and supporting foundation, respectively, to the wavelet coefficients of horizontal seismic accelerations. These transfer functions have the following expressions:

$$\tau_i(\omega) = \frac{K_x / m_f}{\left[\dfrac{K_x}{m_f} - (1+\gamma)\omega^2\right]\left[\left(\omega_i^2 - \omega^2\right) + i(2\zeta\omega_i\omega)\right] - \gamma\omega^4} \tag{7.18}$$

$$\tau_f(\omega) = \frac{\dfrac{K_x}{\omega^2 m_f}\left[\left(\omega_i^2 - \omega^2\right) + i\left(2\zeta\omega_i\omega\right)\right]}{\left[\dfrac{K_x}{m_f} - (1+\gamma)\omega^2\right]\left[\left(\omega_i^2 - \omega^2\right) + i(2\zeta\omega_i\omega)\right] - \gamma\omega^4} \tag{7.19}$$

The expressions relating the wavelet coefficients of base shear and base moment to the wavelet coefficients of the ground acceleration have been derived from Equations (7.14) and (7.15) and may be written respectively as

$$\sum_i W_\psi Q(a_j, b_i)\hat{\psi}_{(a_j,b_i)}(\omega) = \tau_Q(\omega)\sum_i W_\psi\, \ddot{x}_g(a_j, b_i)\hat{\psi}_{(a_j,b_i)}(\omega) \tag{7.20}$$

$$\sum_i W_\psi M(a_j, b_i)\hat{\psi}_{(a_j,b_i)}(\omega) = \tau_M(\omega)\sum_i W_\psi\, \ddot{x}_g(a_j, b_i)\hat{\psi}_{(a_j,b_i)}(\omega) \tag{7.21}$$

The transfer functions $\tau_Q(\omega)$ and $\tau_M(\omega)$ in Equations (7.20) and (7.21), finally, may be expressed as follows (after dividing these functions respectively by m_ig and m_igH to obtain nondimensional coefficients of base shear and base moment):

$$\tau_Q(\omega) = -\frac{1}{g}\left[\frac{\left\{\dfrac{K_x}{m_f}\right\}\left\{\omega_i^2 - i2\zeta\omega_i\omega\right\}}{\left\{\dfrac{K_x}{m_f} - (1+\gamma)\omega^2\right\}\left\{\left(\omega_i^2 - \omega^2\right) + i(2\zeta\omega_i\omega)\right\} - \gamma\omega^4}\right]. \tag{7.22}$$

$$\tau_M(\omega) = -\frac{h_i}{gH}\left[\frac{\left\{\dfrac{K_x}{m_f}\right\}\left\{\omega_i^2 - i2\zeta\omega_i\omega\right\}}{\left\{\dfrac{K_x}{m_f} - (1+\gamma)\omega^2\right\}\left\{\left(\omega_i^2 - \omega^2\right) + i(2\zeta\omega_i\omega)\right\} - \gamma\omega^4}\right] \tag{7.23}$$

Subsequently, using the time-localization property of wavelets, assuming a point-wise relation and taking advantage of narrow energy bands, the instantaneous power spectral density functions for base shear and the overturning base moment may be computed from the transfer functions as obtained in Equations (7.22) and (7.23) as follows:

$$S_{x_i}(\omega) = \sum_j \frac{E\left[\left|X_i^j(\omega)\right|^2\right]}{\Delta b} = \sum_j \frac{K''\Delta b}{a_j} E\left[\left|W_\psi \ddot{x}_g(a_j,b_i)\right|^2\right]\left|\tau(\omega)\right|^2\left|\hat{\psi}_{a_j,b_i}(\omega)\right|^2 \tag{7.24}$$

In Equation (7.24), $\tau(\omega)$ represents either $\tau_Q(\omega)$ or $\tau_M(\omega)$. This expression of PSDF and other statistical estimates would finally give us the expected largest peak responses of the superstructure, which is a tank in this case, in the form of shear and an overturning moment at the tank base. Once these values become known, the same can be applied at the surface of the soil medium modelled with finite elements to obtain the responses with the soil domain. An example is considered here to clarify the approach.

Let us assume a tank resting on the foundation that is placed on the surface of a homogeneous elastic soil medium. The height, radius, wall thickness, elastic modulus, mass density, Poisson's ratio and damping ratio of the tank are assumed as 12 m, 4 m, 4 mm, 2.1×10^8 kN/m^2, 7850 kg/m^3, 0.2 and 5%, respectively. The values of linear elastic soil properties, viz. elastic modulus, mass density and Poisson's ratio, have been assumed as 3.5×10^5 kN/m^2, 1500 kg/m^3 and 0.33, respectively. The tank is filled with water, which has a mass density of 1000 kg/m^3. The values of shear and the overturning moment obtained at tank base from the wavelet-based analysis have been given in Table 7.3. The wavelet-based pseudospectral acceleration (PSA) spectra of the tank have been computed for all three seismic motions at 5% damping and shown in Figure 7.32. It may be seen from this figure that the Miyagi earthquake generated dominant PSA responses for the widest range in a time period (0.06 to 5.0 s), and indeed the strongest of the three motions, as is also evident from the values of the base shear and moment in Table 7.3. The Loma Prieta seismicity has a milder effect on PSA responses and is critical for a shorter range of time (0.1 to 1.0 s). The El Centro ground motion resulted in the mildest peak PSA responses for a range similar to that of Loma Prieta, but without any sharp peaks.

Table 7.3 Expected largest peak value of responses at tank base

Responses	*Loma Prieta (1989)*	*Miyagi (1978)*	*El Centro (1940)*
Base shear (kN)	1207.60	3902.90	446.34
Base moment (kN–m)	661.73	2138.80	244.75

Next, a three-dimensional finite element model of the underlying soil domain can be constructed in any standard software using a linear or non-linear soil constitutive model, and the particular base shear and overturning base moment as specified in Table 7.3 may be applied to the model suitably. The corresponding soil responses may be subsequently obtained from finite element analysis of the model. The model and the approach toward soil–structure–soil interaction analysis are described in the following section.

7.3.2 Finite element model of the system

A three-dimensional (3D) soil-footing model has been developed in DIANA, and a 3D nonlinear static analysis of this model has been carried out separately for each of the three seismic motions considered. The finite element (FE) model is shown in Figure 7.33. The circular tank of radius 4 m is placed at the centre of the soil surface. The soil domain is considered to have a lateral extension that is seven times the tank radius (i.e. 28 m) on both sides along two mutually perpendicular directions. The vertical extent of the soil domain is assumed to be five times the tank radius (i.e. 20 m). In the finite element model, the circular concrete footing is assumed to be 1 m thick. The footing and the soil together have been meshed by 2D plate elements and six-node isoparametric solid wedge elements, respectively. The base shear obtained from the analysis has been distributed equally among all the nodes of the footing along the horizontal direction (+X), and the

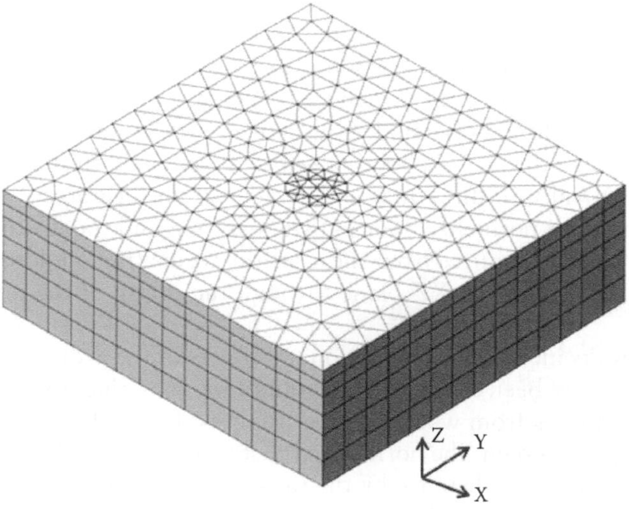

Figure 7.33 3D FE model of the soil in DIANA.

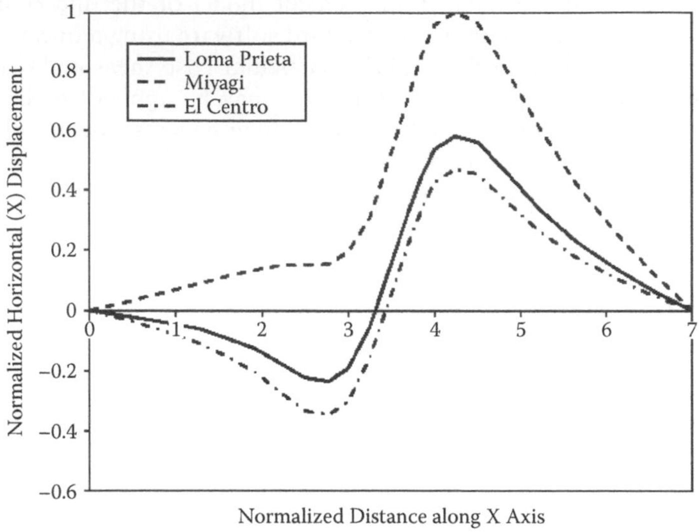

Figure 7.34 Horizontal (X) displacements in soil at 7 m depth. (From DIANA analysis.)

overturning base moment has been equally distributed about the *Y* axis at the nodes at the centre of the footing (at $x = 28$ m). The self-weight of the soil and the tank has been considered during the analysis. The weight of the steel tank in this case (with water) is approximately 5940 kN.

The soil has been represented by Mohr–Coulomb parameters (cohesion = 5 kN/m^2, friction angle = 36° and dilatancy angle = 6°), and $K_0 = 0.5$ is also assumed in addition to the values stated above for linear elastic analysis done before. The elastic modulus, Poisson's ratio and the density of the footing material are 3×10^7 kN/m^2, 0.2 and 2400 kg/m^3, respectively. Figure 7.34 represents normalized horizontal displacements of the soil at a depth of 7 m on the *XZ* plane through the footing centre as obtained from finite element analysis in DIANA for three different seismic motions. Figures 7.35 and 7.36 show variation of normalized normal stresses (S_{xx}) and the maximum horizontal principal strain (free from shear strain) inside the soil domain at 7 m below the surface on the *XZ* plane through the centre of the tank-footing system.

These results include the nonstationary effects of the ground motions, as these effects have been captured while obtaining the values of the shear and moment responses from wavelet-based analysis. It may be mentioned here that in the plots shown, the normalized distance along the *X* axis has been obtained dividing the distance by the tank diameter, and the values of the displacements and stresses have also been normalized, dividing the respective values by the corresponding maximum value obtained considering all

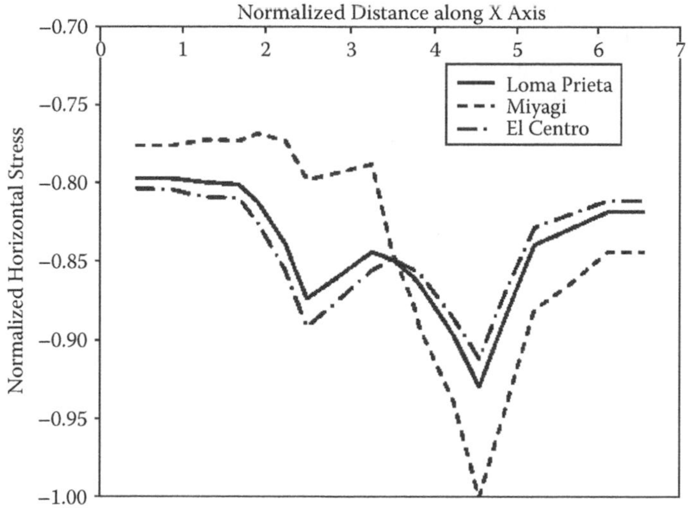

Figure 7.35 Horizontal (*X*) stresses in soil at 7 m depth. (From DIANA analysis.)

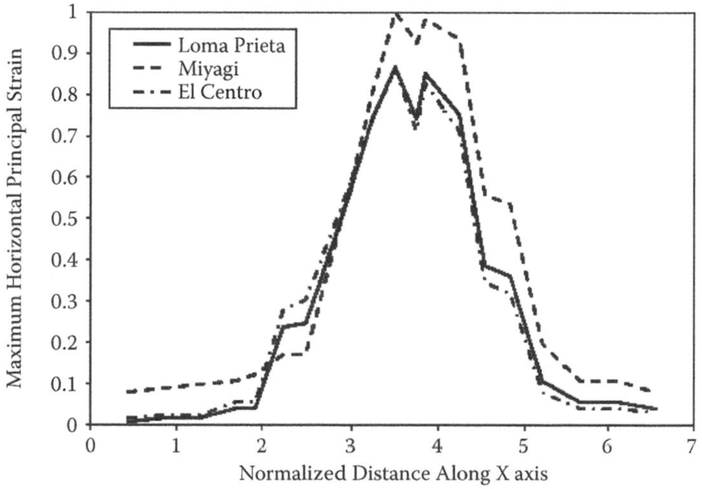

Figure 7.36 Horizontal (*X*) principal strain in soil at 7 m depth. (From DIANA analysis.)

three types of earthquakes. The soil near the boundary is deformed more due to the more powerful Miyagi earthquake, and the nature is antisymmetric in the case of milder earthquakes like Loma Prieta and El Centro (Figure 7.34). As the base shear is applied in the $+X$ direction and the moment has been applied clockwise about the Y axis, so the stresses seem to increase immediately on the right side of the centreline of the tank-soil model (around $x = 3.5$), as evident from Figure 7.35. The maximum horizontal principal strain (Figure 7.36) seems to follow a symmetrical trend about the centreline of the tank.

Appendix: MATLAB® Programmes

```
% Programme File #1: wvcf.m
% This Matlab programme calculates wavelet coefficients of
an input accelerogram based on
% modified Littlewood-Paley basis function (vide Section
2.4) and subsequently characterizes
% the ground motion. The earthquake acceleration time
history data is assumed to be stored in
% file 'acc.dat' in a single column which has 2047 time
points (so 2047 rows).
%                      Author: Pranesh Chatterjee
%                      Year:    2014
%                      Version: 2014.02
%
close all;
clear all;
% User input starts
tsm=0.02;
nn=2047;
lowband=-17;
highband=4;
sigma=2.0^(1.0/4.0);
fid=fopen('acc.dat');  % input acceleration file is read
% User input finishes
tm=(nn-1)*tsm;
zz=fscanf(fid,'%f');
fclose(fid);
for ibk=1:nn  % time instants
    bk=(ibk-1)*tsm;
    for j=lowband:highband % frequency bands
        mj=j+18;
        aj=sigma^j;
        w=0.0;
        for i=1:nn
            ti=(i-1)*tsm;
            if((ti-bk)==0.00)
```

```
                psi=sqrt(sigma-1.0);
            else
                u=(ti-bk)/aj;
                rnum=sin(sigma*pi*u)-sin(pi*u);
                den=u*pi*sqrt(sigma-1.0);
                psi=rnum/den;
            end
                w=w+zz(i)*psi*tsm;
        end
                baj=sqrt(abs(aj));
                wvcfacc(ibk,j+18)=w/baj;
    end
end
save(,wvcfacccmpsps.mat',‘wvcfacc‘); %2047 rows, 22 columns
%
% Calculation of wavelet coefficients finished
%
% Ground motion characterization starts now (vide Section
3.1)
%
for i=1:2047
    for j=1:22
        wvcfaccsq(i,j)=wvcfacc(i,j)^2;
    end
end
        bb=0.2;
        cpsi=2.0*log(sigma)/(2.0*(sigma-1.0)*pi);
        sk=2.0/(4.0*pi*cpsi)*(sigma-1.0/sigma);
        skk=sk/((sigma-1.0)*pi);
        for jj=-17:4
            j=jj+18;
            tj=(4.0*(sigma^jj))/(1.0+sigma);
            jjj=tj/tsm;
            im=1;
            sum=0.0;
            is=0;
            for i=im:2047
                is=is+1;
                sum=sum+wvcfaccsq(i,j);
                if((is<jjj)&&(i==2047))
                    jjj=is;
                    wvs=sum/jjj;
                    ik=i-jjj+1;
                    for ie=ik:i
                        wv(ie,j)=wvs;
                    end
```

```
                is=0;
                sum=0.0;
            elseif(is==jjj)
                wvs=sum/jjj;
                ik=i-jjj+1;
                for ie=ik:i
                    wvcfaccsq(ie,j)=wvs;
                end
                is=0;
                sum=0.0;
            else
                flag=1.0;
            end
        end
    end
    save('wvcfaccsq.mat','wvcfaccsq');  % squared
    acceleration stored
    time_pts=[0.00:0.02:40.92];
    time_pts=time_pts';
    for i=1:2047
        for j=1:22
            tm_inst(i,j)=time_pts(i,1);
        end
    end
    freq_bands=[1:22];
    for i=1:2047
        for j=1:22
            bands(i,j)=freq_bands(1,j);
        end
    end
    %
    % Plotting wavelet coefficients in 3D
      figure(1);
    plot3(tm_inst(:,1:4:22),bands(:,1:4:22)-
    18,wvcfacc(:,1:4:22)/100);
    xlabel(,Time (s)');
    ylabel(,Frequency band');
    zlabel('Wavelet coefficients');
    legend('j = -17','j = -13','j = -9','j = -5','j =
    -1','j = 3');
    %
    % Plotting squared wavelet coefficients in 3D after
      ground motion characterization
   figure(2);
    plot3(tm_inst(:,1:4:22),bands(:,1:4:22)-
    18,wvcfaccsq(:,1:4:22));
```

```
        xlabel('Time (s)');
        ylabel('Frequency band');
        zlabel('Wavelet coefficients squared');
        legend('j = -17','j = -13','j = -9','j = -5','j =
        -1','j = 3');
%     Ground motion characterization is finished
```

```
%     Programme File #2: PSAwavelet.m
%     This program finds out the PSA of a SDOF system based
      on wavelet analytic technique using
%     modified Littlewood Paley basis function (vide Sections
      3.2 & 3.5.2).
%     It reads in squared wavelet coefficients generated by
      previous programme (wvcf.m) and stored
%     separately after ground motion characterization is
      done.
%
%                    Author: Pranesh Chatterjee
%                    Year: 2014
%                    Version: 2014.01
clear all;
close all;
% User input starts here
ze=0.05;       % damping ratio of the SDFOF system
bb=0.2;
grv=9.81;      % acceleration due to gravity (in SI unit)
sig=2^0.25;
dw=0.005;
tsm=0.02;                  % sampling time (in s) in time
                             history acceleration input file
load('wvcfaccsq.mat'); % The squared wavelet coefficients
                           are read
%
cpsi=2.0*log(sig)/(2.0*(sig-1.0)*pi);
sk=2.0/(4.0*pi*cpsi)*(sig-1.0/sig);
skk=sk/((sig-1.0)*pi);
maxtmpt=length(wvcfaccsq);    % length is 2687 for El Centro
wv=wvcfaccsq;
icnt=0;
for itn=0.01:0.025:10.05            % time periods (0.01 s
                                      through 10.05 s)
    icnt=icnt+1;
    wni(icnt)=2*pi/itn;
    timep(icnt)=itn;
end
for ih=1:length(wni)
    for j=-17:4
        sll=pi/(sig^j);
        ul=sll*sig;
        ji=j+18;
        fr(1,ji)=0.0;
        fr(2,ji)=0.0;
        fr(3,ji)=0.0;
        for w=sll:dw:ul
```

```
            fra=1.0/((w*w-wni(ih)*wni(ih))^2.0+(2.0*ze*w*wni
            (ih))^2.0);
            fr(1,ji)=fr(1,ji)+fra*dw;
            fr(2,ji)=fr(2,ji)+fra*w*dw;
            fr(3,ji)=fr(3,ji)+fra*w*w*dw;
        end
    end
    for ii=1:3
        for i=1:maxtmpt
            sm(ii,i)=0.0;
            for j=-17:4
                jj=j+18;
                sm(ii,i)=sm(ii,i)+skk*wv(i,jj)*fr(ii,jj);
            end
        end
    end
    for ii=1:3
        for i=1:maxtmpt
            sm(ii,i)=sm(ii,i)*(wni(ih)*wni(ih)/grv)^2.0;
        end
    end
    for i=1:maxtmpt
        omega(i)=sqrt(sm(3,i)/sm(1,i));
        part1=sm(2,i)^2.0;
        part2=sm(1,i)*sm(3,i);
        part=part1/part2;
        slambda(i)=sqrt(1-part);
    end
    x=0.0005;
    exlpk=0.0;
    prdisfn1=0;
    delta=0.00025;
    ijl=99*(10^6);
    for jk=1:ijl
        alpha=0.0;
        for i=1:maxtmpt
            snulambda=slambda(i)^(1.0+bb);
            alpha1=(omega(i)/pi)*exp(-(x*x/(2.0*sm(1,i))));
            alpha2=1.0-exp(-sqrt(pi/2.0)*snulambda*(x/
            sqrt(sm(1,i))));
            alpha3=1.0-exp(-(x*x/(2.0*sm(1,i))));
            alpha4=alpha1*alpha2/alpha3;
            alpha=alpha+alpha4;
         end
         prdisfn2=exp(-alpha*tsm);
            diff=prdisfn2-prdisfn1;
            if(jk==1)
               exlpk=(x/2.0)*diff+exlpk;
```

```
            else
               exlpk=exlpk+diff*(x-delta/2.0);
            end
            prdisfn1=prdisfn2;
            x=x+delta;
            if(prdisfn2>0.95)
                break
            end
    end
    larpeak(ih)=exlpk;
end
save(,PSAacc.mat','wni','larpeak'); % The largest peak
                                       response of SDOF
                                     % system is saved in a
                                       ,mat' file.
figure(1);
plot(timep,larpeak);                   % Plots PSA spectrum
                                         at 5% damping
xlabel(,Time period (s)');
ylabel(,PSA (in g)');
figure(2);
semilogx(timep,larpeak);           % Semilog plot of PSA
                                     spectrum at 5% damping
xlabel(,Time period (s)');
ylabel(,PSA (in g)');
```

```
% Programme File #3: linstoch.m
% This Matlab programme performs wavelet analysis of a
  soil-tank-footing interaction problem
% where the system is subjected to strong ground shaking
  (vide Section 4.7.1). Values of main
% parameters are:
% Shear wave velocity: vs = 200 m/s
% Tank heights ('ht') considered: ht = 3, 6 and 10 m
% Height-radius ratio of the tank assumed: hr = 3.0
  (slender)
%
clear all;
close all;
%
zei=0.05;                      % damping ratio of impulsive
                                 liquid mass
gr=9.81;                       % acceleration due to gravity
                                 (SI unit)
est=2.1*(10.0^11.0);           % Elastic modulus of steel tank
                                 (N/m^2)
pst=7850.0;                    % mass density of steel tank
                                 material
tsm=0.02;
bb=0.2;
g=9.81;
sig=2^0.25;
dw=0.0001;
%
ep=sqrt(est/pst);
cpsi=2.0*log(sig)/(2.0*(sig-1.0)*pi);
sk=2.0/(4.0*pi*cpsi)*(sig-1.0/sig);
skk=sk/((sig-1.0)*pi);
load(,wvcfaccsq.mat');        % Squared wavelet coefficients
                                are read
wv=wvcfaccsq;                 % Squared wavelet coefficients
                                are stored in a variable
clear wvcfaccsq;
vs=200.0;
ht=[3 6 10];
for ikl=1:3
    for ihr=1:3
      if(ihr==1)
          hr=0.5;
          ci=0.0719;
      elseif(ihr==2)
          hr=1.0;
          ci=0.0875;
```

```
        else
            hr=3.0;
            ci=0.0792;
        end
        rt=ht(ikl)/hr;
        wn=(ci/ht(ikl))*ep;
        wf=62.83; % foundation frequency in rad/s
        wfsq=wf*wf;
        rof=8.0;
        for j=-17:4
             sll=pi/(sig^j);
             ul=sll*sig;
             ji=j+18;
             fr(1,ji)=0.0;
             fr(2,ji)=0.0;
             fr(3,ji)=0.0;
             for w=sll:dw:ul
                 %      IMPULSIVE DISPLACEMENT
                 xa0=w*rt/vs;
                 xa=wfsq;
                 xb=0.65*xa0*xa;
                 xy1=wn*wn-w*w;
                 xy2=2.0*zei*wn*w;
                 xp=xa;
                 xq=xb;
                 xr=(xa-w*w*(1.0+rof))*xy1-xy2*xb-rof*w^4;
                 xs=(xa-w*w*(1.0+rof))*xy2+xy1*xb;
                 fra=(xp*xp+xq*xq)/(xr*xr+xs*xs);
                 fr(1,ji)=fr(1,ji)+fra*dw;
                 fr(2,ji)=fr(2,ji)+fra*w*dw;
                 fr(3,ji)=fr(3,ji)+fra*w*w*dw;
             end
        end
        for ii=1:3
            for i=1:2047
                sm(ii,i)=0.0;
                for j=-17:4
                    jj=j+18;
                    sm(ii,i)=sm(ii,i)+skk*wv(i,jj)*fr(ii,jj);
                end
            end
        end
           for iy=1:2047
               timep(iy)=(iy-1)*0.02;
               xmt(ihr,iy,ikl)=sqrt(sm(1,iy)); % in 'm'
           end
    end
end
```

```
save('linstochres.mat','timep','xmt');
% Results are saved in a 'mat' file
% Results for the tank with height-radius ratio = 3 are now
  plotted
figure(1);
plot(0.0:0.02:40.92,xmt(3,:,1),0.0:0.02:40.92,xmt(3,:,2),0.0
:0.02:40.92,xmt(3,:,3));
%
% Soil-tank-footing interaction analysis is finished
```

References

1. Haar, A. (1910). Zur theorie der orthogonalen Funktionensysteme (Erste Mitteilung). *Math. Ann.*, 69, 331–371.
2. Stromberg, J.O. (1982). A modified Franklin system and higher order spline systems on Rn as unconditional bases for Hardy spaces. In W. Beckner et al. (eds.), *Proceedings of Conference in Honor of A. Zygmund*, Vol. II, Wadsworth Mathematics Series, pp. 475–493.
3. Meyer, Y. (1992). *Wavelets and operators*. Cambridge University Press, New York.
4. Grossmann, A., and Morlet, J. (1984). Decomposition of Hardy functions into square integrable wavelets of constant shape. *SIAM J. Math. Anal.*, 15(4), 723–736.
5. Mallat, S. (1989). Multiresolution approximation and wavelets. *Trans. Am. Math. Soc.*, 315, 69–88.
6. Mallat, S. (1989). A theory for multi-resolution signal decomposition the wavelet representation. *IEEE Pattern Anal. Machine Intell.*, 11(7), 674–693.
7. Daubechies, I. (1996). Where do wavelets come from? *Proc. IEEE*, 84(4), 510–513.
8. Daubechies, I. (1992). *Ten lectures on wavelets*. Society for Industrial Applied Mathematics, Philadelphia, PA, 1992.
9. Newland, D.E. (1993). *An introduction to random vibrations, spectral and wavelet analysis*. Longman, Harlow, Essex, UK.
10. Kaiser, G. (2010). *A friendly guide to wavelets*. Modern Birkhäuser Classics, Birkhäuser, New York.
11. Chui, C.K. (1992). *An introduction to wavelets*. Academic Press, San Diego, CA.
12. Basu, B., and Gupta, V.K. (1998). Seismic response of SDOF systems by wavelet modeling of nonstationary processes. *ASCE J. Eng. Mech.*, 124(10), 1142–1150.
13. Alkemade, J.A.H. (1993). The finite wavelet transform with an application to seismic processing. In T.H. Koornwinder (ed.), *Wavelets: An elementary treatment of theory and applications*. World Scientific Co. Pte. Ltd., Hackensack, NJ, pp. 183–208.
14. Vanmarcke, E.H. (1975). On the distribution of the first-passage time for normal stationary random processes. *ASME J. Appl. Mech.*, 42, 215–220.
15. Gupta, V.K., and Trifunac, M.D. (1993). A note on the effects of ground rocking on the response of buildings during 1989 Loma Prieta earthquake. *Earthquake Eng. Eng. Vibration*, 13(2), 12–28.

16. Basu, B., and Gupta, V.K. (1997). Nonstationary seismic response of MDOF systems by wavelet transform. *Earthquake Eng. Struct. Dyn.*, 6, 1243–1258.
17. Basu, B., and Gupta, V.K. (2000). Stochastic seismic response of single-degree-of freedom systems through wavelets. *Eng. Struct.*, 22, 1714–1722.
18. Wong, H.L., and Trifunac, M.D. (1979). Generation of artificial strong motion accelerograms. *Earthquake Eng. Struct. Dyn.*, 7, 509–527.
19. Lee, V.W., and Trifunac, M.D. (1985). Torsional accelerograms. *Soil Dyn. Earthquake Eng.*, 4(3), 132–139.
20. Lee, V.W., and Trifunac, M.D. (1987). Rocking strong earthquake accelerograms. *Soil Dyn. Earthquake Eng.*, 6(2), 85–89.
21. Lee, V.W., and Trifunac M.D. (1989). A note on filtering strong motion accelerograms to produce response spectra of specified shape and amplitude. *Eur. Earthquake Eng.*, 111 (2), 38–45.
22. Trifunac, M.D. (1990). Curvograms of strong ground motions. *ASCE J. Eng. Mech.*, 116(6), 1426–1432.
23. Housner, G.W. (1963). The dynamic behaviour of water tanks. *Bull. Seismol. Soc. Am.*, 53(2), 381–387.
24. Haroun, M.A. (1983). Vibration studies and tests of liquid storage tanks. *Earthquake Eng. Struct. Dyn.*, 11(2), 179–206.
25. Peek, R. (1988). Analysis of unanchored liquid storage tanks under lateral loads. *Earthquake Eng. Struct. Dyn.*, 16(7), 1087–1100.
26. Peek, R., and Jennings, P.C. (1988). Simplified analysis of unanchored tanks. *Earthquake Eng. Struct. Dyn.*, 16(7), 1073–1085.
27. Veletsos, A.S., and Tang, Y. (1990). Soil–structure interaction effects for laterally excited liquid storage tanks. *Earthquake Eng. Struct. Dyn.*, 19, 473–496.
28. Chatterjee, P., and Basu, B. (2001). Non-stationary seismic response of tanks with soil interaction by wavelets, *Earthquake Eng. Struct. Dyn.*, 30, 1419–1437.
29. Malhotra, P.K. (1997). Seismic response of soil-supported unanchored liquid storage tanks. *ASCE J. Struct. Eng.*, 123(4), 440–450.
30. Veletsos, A.S., and Tang, Y. (1986). Dynamics of vertically excited liquid storage tanks. *ASCE J. Struct. Div.*, 112, 1228–1246.
31. Chatterjee, P., and Basu, B. (2004). Wavelet-based non-stationary seismic rocking response of flexibly supported tanks. *Earthquake Eng. Struct. Dyn.*, 33, 157–181.
32. Veletsos, A.S., and Verbic, B. (1973). Vibrations of visco-elastic foundations. *Earthquake Eng. Struct. Dyn.*, 2, 87–102.
33. Wolf, J.P. (1988). *Soil-structure interaction analysis in time domain*. Prentice-Hall, Englewood Cliffs, NJ.
34. Kondner, R.L. (1963). Hyperbolic stress-strain response: cohesive soils. *ASCE J. Soil Mech. Found Div.*, 89, 115–143.
35. Lambe, T.W., and Whitman, R.V. (1979). Soil mechanics, SI version. Massachusetts Institute of Technology, Cambridge, MA.
36. Das, B.M., and Seeley, G.R. (1975). Load-displacement relationship for vertical anchor plates. *ASCE J. Geotech. Eng.*, 101(GT7), 711–715.
37. Chang, C.Y., Power, M.S., Tang, Y.K., and Mok, C.M. (1989). Evidence of nonlinear soil response during a moderate earthquake. In *Proceedings of 12th International Conference on Soil Mechanical and Foundation Engineering*, Vol. 3, pp. 1927–1930.

38. Roberts, J.B., and Spanos, P.D. (1989). *Random vibration and statistical linearization*. Wiley, New York.
39. Nayfeh, A.H., and Mook, D.T. (1979). *Nonlinear oscillations*. John Wiley, New York.
40. Basu, B., and Gupta, V.K. (1999). Wavelet based analysis of the nonstationary response of a slipping foundation. *J. Sound Vib.*, 222(4), 547–563.
41. Ghanem, R., and Romeo, F. (1999). A wavelet based approach for the identification of linear time-varying dynamical systems. *J. Sound Vibration*, 234(4), 555–576.
42. Ghanem, R., and Romeo, F. (2001). A wavelet based approach for model and parameter identification of nonlinear systems. *Int. J. Nonlinear Mech.*, 36(5), 835–859.
43. Chatterjee, P., and Basu, B. (2006). Nonstationary seismic response of a tank on bilinear hysteretic soil using wavelet transform. *Probabilistic Eng. Mech.*, 21, 54–63.
44. Chatterjee, P., Obrien, E., Li, Y., and Gonzalez, A. (2006). Wavelet domain analysis for identification of vehicle axles from bridge measurements. *Comput. Struct.*, 84, 1792–1801
45. Basu, B. (2005). Identification of stiffness degradation in structures using wavelet analysis. *Constr. Building Mater.*, 19, 713–721.
46. Livaoglu, R. (2008). Investigation of seismic behavior of fluid–rectangular tank–soil/foundation systems in frequency domain. *Soil Dyn. Earthquake Eng.*, 28(2), 132–146.
47. Tam, L. (2008). Seismic response of liquid storage tanks incorporating soil structure interaction. *ASCE J. Geotech. Geoenviron. Eng.*, 134(12), 1804–1814.
48. Ding, W., and Sun, Z. (2014). Wavelet-based response of computation for base-isolated structure under seismic excitation. *J. Appl. Math. Phys.*, 2, 163–169.
49. Ovanesova, A.V., and Suárez, L.E. (2004). Applications of wavelet transform to damage detection in frame structures. *Eng. Struct.*, 26, 39–49.

Index